Electrolytic Separation

Recovery & Refining of Metals

Theoretical and Practical

G. Gore, LL.D., F.R.S.,

ISBN 1-929148-34-8

Wexford College Press
2003

Printed and Published by
"THE ELECTRICIAN" PRINTING AND PUBLISHING CO., LIMITED,
1, 2, and 3, Salisbury Court, Fleet Street,
London, E.C.

chemistry, thermo-chemistry, voltaic action, and electrolysis, very expensive experiments upon a large scale have been made in electrolytic refineries, especially the earliest established ones, in order to determine the conditions of practical and commercial success, under which the process could be conducted in the most economical manner and with pecuniary profit, more particularly as regards the most suitable number and dimensions of vats. Great expense has also been incurred by makers of dynamos in improving those machines, and by electrolytic refiners in substituting one kind of dynamo for another in order to obtain the most suitable.* As nearly every inventor wishes his invention to be quickly published and used ; nearly as fast as those improvements in refining and in dynamos have been made, information respecting them has been published in various periodicals, and all that is suitable of it has been condensed and arranged in a systematic order in this work.

The present book contains both the Science and the Art of the subject, *i.e.*, both the Theoretical Principles upon which the art is based, and the Practical Rules and details of technical application on a commercial scale, in order thereby to render it suitable both for students and manufacturers. Whilst the fundamental principles of the art are fully described, it is not intended to supply a treatise containing information respecting chemistry of the metals, because the reader is supposed to be already acquainted with that subject as a necessary preliminary to enable him to understand electrolytic processes. Nor is it intended to furnish full information respecting dynamo-electric machines, because that is

---

* More than thirty dynamos have been thrown aside for improved ones in a single refinery.

largely a separate and preliminary matter; sufficient, however, is said respecting the chemical relations of electrolysis, and respecting dynamos, to afford general guidance in the art of electrolytic refining. Only those portions of the subjects are included, a knowledge of which is indispensable to the successful working of the process. The special subjects required to be previously known in order to fully understand this one are chemistry, voltaic electricity, magnetism, and electrical measurement; the necessary preliminary knowledge of these the reader is supposed to already possess.

As the process has not yet been extended as a commercial success greatly beyond the separation and refining of copper, not much can be said respecting its attempted application to other metals; what has been done on a commercially successful scale is, however, briefly described. Many commercially unsuccessful processes are omitted.

The subject of the book is treated throughout in a thoroughly systematic manner. First is given a brief historical sketch of the origin and development of the sciences of voltaic and magneto electricity, the dynamo-electric machine, and of electrolysis and its application to the refining of copper. Next comes the Theoretical Division, treating of, in succession in separate sections, the chief electrical, thermal, chemical, voltaic, electrolytic, magneto and dynamo electric facts and principles of the subject. Of these sections the one on electrolysis is the most full and complete, omitting no known truth of importance relating to the subject; it also includes some useful tables of the rates of corrosion of various metals at different temperatures, the amounts of electric energy and horse-power expended in electrolysing various substances, the amount lost in heating the conductors, &c. Then follows the Practical Division, containing valuable information respecting the mode of establishing an electrolytic refinery, the amount of space necessary, the number and size of the vats, the amount of electric energy and horse-power required; the preparation of the electrolyte, arrangement of the vats, electrodes and main conductors ; the mode of conducting the process, &c., so as to obtain pure copper at the least cost ; brief descriptions are also given of various processes adopted or attempted for refining other metals than copper, and for operating upon cupreous minerals without previously smelting them ; also for recovering gold from auriferous earths ; concluding with an account of Cowles's electric smelting furnace ; and an appendix of data, &c., useful in electrolytic refining.

As various alterations and improvements are continually being made in dynamo-electric machines, in electrolytic processes, &c., in electrolytic refining works, and accounts of these alterations are not always immediately published, some of the statements given necessarily represent the circumstances that existed a short time ago ; nevertheless, no trouble has been spared to bring the descriptions generally well up to date. Some of the previously published statements, especially the numerical ones, are so inaccurate that they have had to be omitted, but all that could be have been corrected and inserted. Many of the data have been obtained direct from electrolytic refiners and dynamo makers.

Several points of useful information respecting dynamo-electric machines the author has obtained from the valuable works on " Dynamo-electric Machinery," by S. P. Thompson, and " Magneto and

Dynamo Electric Machines," by W. B. Esson. To various manufacturers of those machines also who have afforded him information respecting their own particular kinds of dynamo, and to proprietors of electrolytic refineries who have freely answered his enquiries, he has to offer his thanks. To several correspondents in America he is especially indebted for information of the state of the art in that country.

A number of points of practical information the Author has extracted from " Electrolyse," by M. Fontaine ; M. Kiliani's Paper in the *Berg und Hüttenmännische Zeitung*, 1885, and M. Badia's Papers in *La Lumière Électrique*, 1884 ; also from numerous articles in *The Electrician*, Dingler's *Polytechnisches Journal*, the *Proceedings* of the Institute of Civil Engineers, *Industries*, *Engineering*, *The Scientific American Supplement*, *The Electrical Review*, &c.

The Author has also to acknowledge the loan of several of the illustrations appearing in the book.

CONTENTS.

———◆———

———

1. THEORETICAL DIVISION.

SECTION A.

SECTION A. (continued).

SECTION B.

SECTION C.

SECTION D.

SECTION D. (*continued*).

SECTION E.

SECTION E. (continued).

SECTION E. (continued).

SECTION F.

2. PRACTICAL DIVISION.

SECTION G.

SECTION G. (continued).

SECTION G. (continued).

SECTION H.

OTHER APPLICATIONS OF ELECTROLYSIS IN SEPARATING AND REFINING METALS.

SECTION H. (continued).

APPENDIX.

ELECTROLYTIC SEPARATION

AND

REFINING OF METALS.

HISTORICAL SKETCH.

More than thirteen hundred years ago Zosimus mentioned the earliest known fact respecting the electrolytic separation of metals, viz., that by immersing a piece of iron in a cupreous solution it acquired a coating of copper. Ever since that time the same fact has been commonly observed by workers in copper mines, that their tools of iron or steel became coated with a film of copper by contact with the water percolating through the mines ; the water holding in solution blue vitriol derived from oxidation of mineral sulphides of copper contained in the rocks. Paracelsus in the years 1493 to 1541, and even Stisser, the professor of chemistry in Helmstadt, as late as 1690, believed that by this process the iron was changed into copper.

The discovery of chemical electricity by Volta and the invention of the voltaic battery as an instrument for producing it did not occur until about the year 1799. Wollaston soon afterwards observed that " if a piece of silver in connection with a more positive metal be put into a solution of copper, the silver is coated over with copper, which coating will stand the operation of burnishing" (*Philosophical Transactions* of the Royal Society, 1801). About the same time Mr. Cruickshank

B

passed an electric current from his voltaic battery through a solution of sulphate of copper, and found that the copper attached itself to the wire connected with the zinc end of the battery, and stated that the metal was "revived completely" (Wilkinson's "Elements of Galvanism," Vol. II., 1804, p. 54). In 1805 Brugnatelli also observed that when the current entered the liquid by means of a piece of copper, the copper was dissolved and then deposited upon the negative pole ("Annals of Chemistry").

In 1831 Faraday discovered magneto-electricity, or the production of electric currents by means of mechanical power acting through the medium of magnets, a discovery which enabled all the subsequent inventions and improvements in dynamo-electric machines to be made, and the refining of metals by electrolysis to become commercially possible on a large scale. When he made this discovery the electric current he obtained was so very feeble that he was barely able to detect it, and he remarked : "I have rather, however, been desirous of discovering new facts and new relations dependent on magneto-electric induction than of exalting the force of those already obtained, being assured that the latter would find their full development hereafter" ("Experimental Researches," para. 158). This prediction has since been abundantly verified.

In 1836 De la Rue observed that the copper deposited by the voltaic current in a Daniell's battery cell gradually became thicker, and might be stripped off in the form of a separate sheet of metal from the surface upon which it had been deposited. About the year 1839 Jacobi, of St. Petersburg, and soon afterwards Jordan, Spencer, and others, made and published their experiments on electrotyping in copper, and thus made the process of depositing that metal familiar to the public.

From that time until the present, copper has been constantly deposited on a commercial scale as a coating upon various articles of iron, &c., in order to protect them ; the electrolytic process has also gradually extended and been employed to form ornaments and other articles, until at length copper of more than one inch in thickness has been deposited, and copper statues weighing several tons have been formed by the process. Electro-deposited copper has also been frequently analysed and found to be extremely pure—so much so, that it has been employed in the Royal Mint to alloy with gold in making the standard coins of this realm. Its deposition on a large scale and great degree of purity, thus foreshadowed the electrolytic refining process.

The first actual commercial application of electrolysis to the refining of copper is to be found in a patent (No. 2,838) granted to James B. Elkington (son of the late G. R. Elkington, the original patentee of commercial electro-silvering and gilding), Nov. 3rd, 1865, entitled " Manufacture of Copper from Copper Ore." In this process plates of crude copper are used as anodes, suspended in " troughs charged with a nearly saturated solution of sulphate of copper," the cathodes or negative plates being formed " of pure copper, rolled very thin." As the crude copper dissolves, pure copper is deposited upon the negative sheets. The patentee proposed to use a series of depositing troughs, each containing a set of properly connected anodes and cathodes, the electrolysis being effected by a current obtained from a magneto-electric machine. The insoluble residue which falls from the anodes to the bottom of the liquid frequently contains " silver, some gold, and also tin and antimony."

In a second patent (No. 3,120), taken out by the same patentee, Oct. 27th, 1869, for the "manufacture of copper, and separating other metals therefrom," impure copper, and especially that which contains much silver, is cast with a " T-shaped head of wrought copper," to enable it to be conveniently suspended as a dissolving plate in the depositing vessel.

"These plates are placed in fireclay jars, ranged longitudinally in troughs on a slightly-inclined, pitched, and otherwise prepared wooden floor, in the dissolving-house. Each jar has a hole in the bottom, closed by a wooden plug, and two holes in the sides, one low down and the other on the opposite side near the top, each jar being connected with the next lower one by a pipe or tube passing from the higher hole of the one to the lower hole of the other jar. The liquid current thus established between the solutions effects mixture of the layers of different density, maintaining all the liquids in the series of jars practically alike at top and bottom, notwithstanding the disturbing influence of the electro-deposition, which constantly tends to produce inequalities of density.

"The solution made use of is water saturated with sulphate of copper, a store of which is kept in a tank at the upper part of the depositing-room, whence it is admitted into the uppermost jars, and runs from jar to jar till all are filled. A solution may also be used obtained by boiling the deposit formed in the culvert or long flue by which the smoke from the copper furnaces is led to the high chimney.

"The tubes connecting the jars have clips attached to their india-rubber portions, acting as stop-cocks. When needful the clips are removed, so as to cause the solutions to mix, the dense layer from the bottom of one jar displacing the lighter portion from the top of another, until the density throughout becomes equalised. The outflow is received in a tank at the lower end of the room, from which it is pumped back to the upper reservoir. The plugged holes at the bottom of the jars enable the latter to be emptied on to the inclined floor, the liquid then flowing into the lower tank. The T-shaped heads attached to the plates enable them to be suspended in the jar, from horizontal copper bars having forks upon them. Interposed between the copper plates to be dissolved are suitably-suspended receiving plates for the deposit of the electrolytic copper. The receiving plates of one jar are connected by a conducting strip of copper with the cast plates of the next jar, and so on throughout. The series being coupled up into a circuit, the terminals are connected with one or more magneto-electric machines. The silver originally contained in the copper of the plates set for solution separates as a sediment, and ultimately accumulates as a deposit in the tank below, after the repeated workings and emptyings of the jars.

"Ores rich in silver are preferred as the source of cast copper for this treatment, since the silver is obtained as a bye product, without increasing the cost of obtaining pure copper from the impure metal."

These two patents of Mr. James Elkington contain the essential parts of the process of purifying copper by means of electrolysis, viz., employing slabs or thick plates of the crude metal as anodes ; a series of depositing vessels with the solution flowing slowly through the whole of them, in order to keep it uniform in composition ; the use of electric currents generated by means of mechanical energy, and the collection of the valuable impurities in the form of a sediment at the bottom of the vessels. The only other essential circumstance remaining is that of the occasional purification of the electrolyte by evaporating it, and crystallising out of it the sulphate of iron and other soluble salts which gradually accumulate.

This process was developed and carried out on a large scale at Pembrey, near Swansea. The works at Pembrey formerly belonged to Messrs. Elkington, Mason and Co., but have since passed into the possession of the Elliott Metal Company (Limited), Selly Oak, near Birmingham, and at the present time this refinery is one of the largest and most perfect existing.

Meanwhile, *i.e.*, during the gradual extension of electrodeposition of copper, the originally minute, but, nevertheless, extremely important fact and phenomenon, of the production of an electric current by mechanical energy acting through the medium of magnetism, was developing slowly. This phenomenon when first discovered by Faraday in 1831 was so small as scarcely to be perceptible, and was first observed by him as "a sudden and very slight effect at the galvanometer" (Faraday's "Experimental Researches," Vol. I., p. 3). No sooner had Faraday published this seemingly unimportant effect than various experimentalists endeavoured to obtain it upon a larger scale. Prof. Forbes, of Edinburgh, Nobili, Ritchie, and others, as well as Faraday himself, quickly succeeded in obtaining electric sparks by means of the magneto-electric current. In 1832 H. Pixii invented and exhibited at Paris the first magneto-electric machine, and decomposed water by means of the current from it; this was followed in 1833 by Saxton's improved machine, and by that of Clarke in 1836 ; and on August 1st, 1842, was granted to J. S. Woolrich the first patent, No. 9,431, for a magneto-electric machine for commercial purposes. Woolrich's machine was for a long time used by Messrs. Prime, of Birmingham, for electro-silverplating.

Since that time the improvements in producing magneto-electricity have been numerous. In 1857 Werner Siemens invented his shuttle-wound armature; in 1860 Pacinotti developed his ring armature machine for yielding a continuous current, which formed the basis of Gramme's and other direct-current machines. In 1866 H. Wilde further increased the power of the apparatus by using a soft iron electro-magnet instead of an ordinary steel one to produce the currents; in 1867 Siemens, Wheatstone and Ladd each separately observed the fact of self-excitation by mechanical power, and developed the self-exciting machine; and in 1871 M. Gramme produced the first practical continuous-current machine for commercial purposes. Alterations and improvements have since that time succeeded each other so rapidly that there exists now quite a large variety of machines for converting mechanical energy into electric current. The magnitude, weight, and speed of the machines have also increased, until there are some which require several hundred horse-power to drive them; others which weigh as much as forty-five tons each, and some with armatures revolving at a speed of nine thousand revolutions a minute. The efficiency of the machine has also been increased until as much as 96 per cent. of mechanical energy imparted to the armature is converted into electric power. With a single dynamo as much as thirty tons of copper are now deposited per week. Even now the limit has not been attained, or even conceived, to the magnitude and uses of the machine, and dynamos will probably yet be constructed equal in amount of energy to the largest steam engines, and their power be applied to a great variety of purposes, mechanical, thermal, and chemical.*

At the present time the process of electrolytic separation and refining of metals is extending rapidly; it is carried on at Pembrey, Widnes, Swansea, Tyldesley (Lancashire), St. Helens (Lancashire), Milton (near Stoke-upon-Trent), Paris, Marseilles, St. Denis, Angoulême, Biache (Pas de Calais), Froges (near Grenoble), Hamburg, Stolberg, Berlin, Moabit (near

---

* Dynamos of "10,000 electric horse-power each," weighing "500 tons," and having armatures "45 feet over all," are being constructed (see *The Electrician*, Vol. XXI., p. 787).

Berlin), Oker (in the Hartz), Eisleben, Burbach (near Siegen), Frankfort-on-the-Maine, Schaffhausen, Stattbergerhütte (near Cologne), the Koenigshütte (in Silesia), Witkowitz (in Moravia), Stephanshütte (in Upper Hungary), Brixlegg (in the Tyrol), Ponte St. Martino (in Piedmont), Casarza (near Genoa), Pittsburg (Pennsylvania), Milwaukee (Wisconsin), Bridgeport (Connecticut), Omaha (Nebraska), Ansonia (Connecticut), St. Louis (Missouri), Newark (New Jersey), Cleveland (Ohio), Longport (near New York), Santiago (Chili), Chihuahua (Mexico), &c. At Messrs. Balbach's Works, Newark (New Jersey), sixty tons of copper are deposited and refined per week. "Messrs. Bolton, at Widnes, and Messrs. Vivian, as well as Messrs. Lambert, at Swansea, are each depositing from forty to fifty tons per week, by currents from 5,000 to 10,000 amperes" (*Nature*, January 26th, 1888, p. 303). The "daily quantity refined at Oker is two and a-half tons ; and the total amount in Germany and Austria is about six tons daily." Messrs. Elliott and Co., of Pembrey (near Swansea), deposit nearly the largest amount, forty-five to sixty tons per week, in this country. I am informed from a direct source that the Bridgeport Copper Company, of Bridgeport (Connecticut), electrolytically refine " one million pounds of copper per month," or about 108 tons per week, "by means of three dynamos."

THEORETICAL DIVISION.

CHIEF FACTS AND PRINCIPLES UPON WHICH THE PROCESS IS BASED.

THE very foundation of electrolysis and dynamo-electric action consists of the principles of energy, conservation of energy, transmutation, correlation and equivalency of different forms of energy, and indestructibility of matter and motion.

Energy may be defined as motion, and that of any given substance or system of bodies is the total amount of motive power which it possesses or can impart. It is either potential or active; potential when stored up in an unchanging condition, ready to produce, but not producing, any dynamic effect, and active when producing some change. Energy may be changed in form, but not altered in total amount, *i.e.*, not created, nor destroyed; mechanical energy may be transformed into heat, electric current, &c., equivalent in amount to that which disappears; heat may be converted into electric current, chemical action, and so on. Work done is resistance overcome, and is often attended by transformation of energy.

Every act and change which occurs in the dynamo-electric machine and in the electrolysis vat involves energy and change of form of energy. The mechanical power of the motor driving the machine disappears, whilst electric current appears in the conducting circuit; and the amount of the former expended is proportionate to that of the latter which is produced, the one being in a large degree equivalent to the other. If the conducting circuit is broken, so that no current passes, the amount of mechanical power consumed to drive the dynamo is small, and is only about equal to that required to overcome the friction of the moving parts, but directly the circuit is completed, and a copious current is generated, the mechanical energy consumed is very great. The total effect produced is in all cases equivalent to the amount of energy expended. In the electrolysis vessel the electric current produces chemical action, and the amount of potential chemical energy of the substances separated in that vessel is equivalent to that of electrical energy required to separate them.

The following are the fundamental units upon which all calculations of energy expended and effects produced (or work performed) in all dynamo-electric and electrolytic actions are based :—

FUNDAMENTAL UNITS OF QUANTITY.

Length = 1·0 centimetre (= ·3937 inch).
Mass = 1·0 gramme (= 15·432 grains).
Time = 1·0 second.

The centimetre-gramme-second (or C.-G.-S.) unit of force is called a DYNE, and is the force which, acting on a mass weighing one gramme during one second, produces in it a velocity of one centimetre per second. The force of one dyne is nearly equal to a weight of 1·02 milligramme on the earth's surface.

The C.-G.-S. unit of work is called an ERG, and is the amount of work done by one dyne acting through a distance of one centimetre, *i.e.*, it is the product of the unit of force by the unit of length. The C.-G.-S. unit rate of working, or unit of power, is one erg per second ; power is the rate of working, and not the amount of work done. Both the dyne and the erg are quantities much too small for ordinary use in technical calculations. For the sake of greater convenience, therefore, what are called " practical units," derived from and quantitatively related to the C.-G.-S. or absolute ones, are usually employed. The practical units are given in the Appendix, for greater convenience of reference.

CHIEF PHENOMENA IN THE ELECTROLYTIC SEPARATION AND REFINING OF METALS.

Some of the chief subjects involved in the electrolytic separation and purification of metals by means of dynamo-electric currents are, electric conduction and insulation; electric-conduction resistance; decomposability of electrolytes, polarisation of electrodes; resistance at electrodes; chemical action; thermo-chemical action; atomic and molecular weights, valency;

chemico-electric or voltaic action; definite electro-chemical change ; magnetic induction ; magneto-electric action ; electro-magnetic action ; electro-dynamic induction, &c.

Conduction and insulation are involved, because the current has to pass through the electric generator, the conductors, the plates, and solutions in the vats, and must be prevented leaking or passing in wrong directions. Conduction resistance, and the heat due to it, have to be considered, because of the attendant waste of mechanical and electric energy. The fact that the same amount of current is able to separate and purify the metal in the solutions of either a few or a large number of vats at the same time, without the expenditure of an increased amount of energy, depends upon the laws of conduction resistance. The decomposability of electrolytes is also a matter of importance, because a difference in its degree involves a change in the amount of power required to be expended. Thermo-chemical action and thermo-chemical equivalents also throw great light upon the amounts of energy necessary to decompose different substances. Upon the atomic weights and chemical valencies of the elements depend the quantities of different metals which a given amount of current will separate and purify. The molecular weights of the acids in the solutions affect the quantity of metal with which they will combine. Upon the law of definite electro-chemical action depends the circumstance that the amount of metal deposited by the same electric current is the same in each of any number of vessels placed in successive connection. Chemico-electric action affects the process, by generating voltaic currents, which may assist or oppose the working current. And magnetic and magneto-electric, electro-magnetic and electro-dynamic action take place in all the different kinds of dynamo-electric machines now employed for the electrolytic separation of metals. All these chief facts and principles, therefore, require to be sufficiently known and understood by every person engaged in superintending such processes.

SECTION A.

CHIEF ELECTRICAL FACTS AND PRINCIPLES OF THE SUBJECT.

Electric Polarity and Induction.—Electric polarity is an electro-static condition of a body which is electrified, and upon

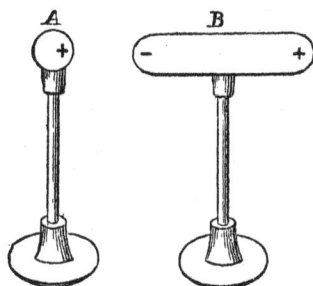

FIG. 1.—Polarity and Induction.

the surface of which a charge of electricity exists; the kind of charge and of polarity being either positive or negative, plus or minus. In consequence of the self-repellent power of electricity, a charged body is in a state of electric tension or potential tending to discharge and assume the electric state of the earth; the latter is assumed to be at zero. Electric induction is an action of an electrified body upon neighbouring ones through a non-conducting or dielectric medium. Electric polarity tends to produce by induction an opposite state of charge and polarity upon the surfaces of all neighbouring bodies. Thus the insulated charged metallic ball, A (Fig. 1), induces a negatively charged state upon the nearest end of the insulated metallic cylinder B. Polarity and induction precede conduction at the first moment of flow or passage of a current.

Electric Quantity.—Electric quantity is either static or dynamic, *i.e.*, quantity of electricity in a state of rest or of motion. The C.-G.-S. (or centigrade-gramme-second) electro-static unit of quantity is that amount of accumulated electricity which at a distance of one centimetre repels an equal quantity of similar electricity with a force of one dyne. The C.-G.-S. unit of quantity is the amount of electricity which is conveyed by unit current in unit time—*i.e.*, in one second; it is about 3×10^{10} times the electro-static unit. The practical unit of quantity of electricity is termed a COULOMB (formerly a weber), and is equal to one-tenth of the C.-G.-S. electro-magnetic unit of quantity.

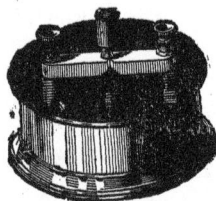

FIG. 2.—Condenser.

Electric Capacity.—The electric capacity of a body is the quantity of electricity it can hold when charged to one unit of potential. This amount varies with the magnitude of the body and with its shape. The best form to retain a charge is that of a smooth sphere, and the worst is that of an elongated body terminated by points. The C.-G.-S. unit of capacity of a conducting body is that which requires a charge of one unit of static electricity to raise it from an uncharged state to one unit of potential, and is equal to that of a smooth metal sphere of one centimetre radius. The capacity of a condenser which holds one coulomb when charged to unit potential of one volt is called a FARAD; but the practical unit of capacity employed is one-millionth part of this, and is termed a MICROFARAD. Commonly a third of a microfarad is used as the most convenient. A condenser has a capacity of one farad when a difference of potential of one volt between its two sets of plates charges each set with one coulomb. According to the single fluid theory of electricity, a condenser is not a store of electric energy, because in the act of charging as much electricity is taken out of one set of its plates as is put into the other.

A condenser (*see* Fig. 2) is usually formed of a pile of alternate circular sheets of tinfoil and of mica or paraffined tissue paper very carefully insulated, every alternate sheet of foil being connected with one terminal of the instrument, and every other alternate sheet with the other, and is provided with a brass plug for connecting the two sets together when required to discharge the instrument by enabling its two electricities to unite.

Electric Potential is electric capability of doing work. We call that condition electric potential which isolated bodies are in when electrically charged, and which in some sense may be likened to the expansive tendency of gases, the hydrostatic pressure of liquids, or the temperature of substances. It is in consequence of its potential that free accumulated electricity tends to discharge and diffuse into electrically neutral bodies and into those of lower potential. As electricity is not known separate from matter, it is more correct to speak of degree of electrification than of amount of electric charge. Electrostatic potential at any point is defined to be "the work that must be expended upon a unit of positive electricity in bringing it up to that point from an infinite distance." Potential (or difference of potential) has also been defined as "the quotient of quantity of electricity by distance." Two bodies having the same positive or the same negative potential produce no current when connected together by a wire. The potential of a conductor depends partly upon the quantity of its electric charge and partly upon the size and shape of the body; the potential at all parts of a charged isolated conducting sphere on which electricity is at rest is equal. With a non-spherical charged conductor the density of the charge and the tendency to discharge are greatest at the parts of greatest diameter and at the most sharp-pointed projections.

Potential differences may be measured by weighing. The practical unit of difference of potential and of electromotive force is termed a VOLT. According to Exner, the earth is at a negative potential of 4·1 volts, ready to discharge its electricity with that degree of pressure into an electrically neutral body; it is, however, usually treated as if it was at zero. The potential of an isolated electrified body, but not the amount of charge, varies on the approach of another body.

Electromotive Force is that power which produces current or which moves, or tends to move, electricity from one place to another, and is, in some sense, analogous to pressure. Difference of electromotive force is considered to be due to difference of potential, and varies with it. That of a current

FIG. 3.—Tripod Galvanometer.

generator is equal to the degree of inequality of potentials at its poles when the latter are disconnected and no current is circulating. A unit of electromotive force (or of potential) exists between two points when one erg of work has to be done in order to urge one unit of positive electricity from the point which is at the lower to that which is at the higher potential. The practical unit of electromotive force, termed a *volt*, is equal to ·9268 of that of a Daniell cell, and is the difference of potential which must be maintained at the ends of a conductor having a resistance of one ohm in order that a current of one ampere may flow through it.

Measurement of Electromotive Force.—There are various ways of effecting this object. It may be done with the aid of an electric condenser, with a suitable galvanometer, a quadrant electrometer, a standard voltaic cell, a thermo-electric pile, a voltmeter, &c.

By successively and separately charging a condenser (*see* p. 12) from two generators of current, and separately discharging it successively through a suitable galvanometer and observing the amounts of swing of the needle, the relative electromotive forces of the two generators may be easily and quickly

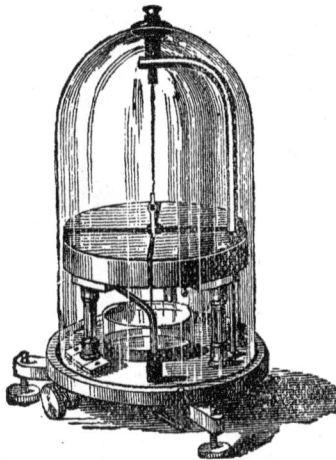

Fɪɢ. 4.—Quadrant Electrometer.

determined; a Thomson's reflecting galvanometer (Fig. 3) of high resistance may be employed for the purpose.

By connecting the terminals of two generators in succession with the poles of a quadrant electrometer (Fig. 4) and observing the amounts of steady deflection of the needle. When the electromotive force is thus measured directly as a difference of potential the circuit is never closed, and no current passes. The difference of potential at the ends of a current-producer of *constant* electromotive force, when no current is circulating, is equal to the electromotive force, and is very nearly so when a very small current is passing

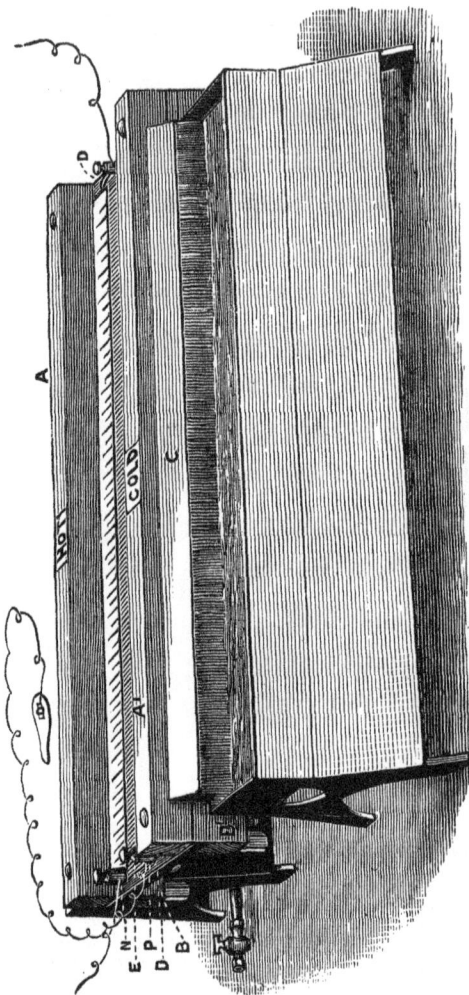

FIG. 5.—Thermo-Electric Pile.

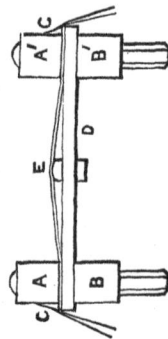

FIG. 6.—Section of Pile.

The relative electromotive forces of the currents from generators of *unchanging* electromotive force may be measured by ascertaining the amounts of resistance through which they respectively force equal strengths of current. In using this method the law of the galvanometer employed need not be known. The relative electromotive forces of currents from two such sources may also be found, either (1) by connecting them together in series, with a suitable galvanometer in the circuit, so that their currents are in the same direction and assist each other ; or (2) so that they oppose each other. We may oppose the current from a generator to that of such a number of standard voltaic cells in single series that no current passes, as shown by a galvanometer in the circuit; the electromotive forces of the two sources are then equal. In using this method neither the resistances of the generators nor that of the galvanometer need be known. As standard cells we may employ either Clark's, the electromotive force of which is equal to 1·438 legal volts, or Daniell's constant battery, a single cell of which has an electromotive force of 1·074 volt (*see* pp. 59—61). If the electromotive force of the current to be measured is constant, and does not much exceed one volt, we may oppose to it that of a current from a thermo-electric pile of many pairs of iron and German silver wires (Figs. 5 and 6) (see *Proceedings* of the Birmingham Philosophical Society, Vol. IV., Part I., p. 130; *The Electrician*, March 15, 1884, Vol. XII., p. 414).* By this method, using a small magnesium-platinum couple in distilled water, I have detected the change of electromotive force caused by adding one part of chlorine to 500,000 million parts of water (*Proceedings* of the Royal Society, Vol. XLIV., 1888, p. 151). One advantage of the method of opposition is that it is a *null* one, being made when no current is passing in either generator, and therefore no polarisation is produced ; it is also a delicate method if the galvanometer employed is a sensitive one. For details of the methods of measurement the reader is referred to special works on the subject.†

---

* *The Electrician* is a weekly illustrated journal devoted to the Practical and Theoretical Applications of Electricity, and is published at 1, 2, and 3, Salisbury Court, Fleet Street, London. The pile referred to is manufactured by Messrs. Nalder Bros. and Co., of Horseferry Road, London.

† Kempe's " Handbook of Electric Testing," Ayrton's " Practical Electricity," Latimer Clark's " Electric Measurement." All books on Electrical and allied subjects can be obtained at *The Electrician* Office.

Electric Current.—An electric current is said to be a flow of electricity from one point to another, but we do not actually know whether it is a flow or not, nor what its real direction is ; it has, however, no existence in the absence of a conductor. Its direction in a voltaic cell is said to be from the zinc through the liquid to the copper, and in an electrolytic one, from the metal which dissolves or evolves oxygen to that which receives a deposit or evolves hydrogen. According to modern theory, the transfer of electric energy is not along the conductor at all, but in the surrounding dielectric, the presence of the conductor, however, being a necessary condition of the transference.

An electric current differs in several fundamental respects from accumulated electricity or an electric charge. Whilst a charge resides chiefly on the surface, a steady current permeates the mass of a substance. A current possesses direction and exhibits magnetism, but electricity at rest does neither ; a charged body, however, when rapidly rotated in a circle is magnetic, like a circular electric current. An electric current, however small, heats a conductor, and decomposes an electrolyte, but a static charge of electricity does not.

The practical unit of quantity of current is called a COULOMB, and is the amount of electricity which passes through a conductor in one second when the strength of current is one ampere, or that which electrolytically deposits ·01725 grain of silver.

Strength of Current.—This is the quantity of electricity which flows through any cross-section of a circuit in one second of time. It has been frequently called "intensity " of current; it depends upon the electromotive force and the total resistance. By using a sufficient electromotive force, and an adequate quantity of electricity, any strength of current may be sent through a conductor until the latter melts or volatilises. No difference has hitherto been proved to exist between any two currents of equal strength. The practical unit of strength of current is called an AMPERE (formerly "weber per second"), and is produced when an electromotive force of one volt acts through a resistance of one ohm, and conveys one coulomb in one second. It is equal to one-tenth of the C.-G.-S., or absolute unit of quantity. A milliampere is one-thousandth of an ampere. According to Ohm's law, the strength of a current is equal to the electromotive force divided by the resistance—

$$S = \frac{E.M.F.}{R}.$$

If a current is passing through a wire, and the resistance remains the same but the electromotive force varies, the strength of current or number of amperes varies directly with that of volts; a galvanometer, therefore, which measures strength of current may be used to measure electromotive force. A current can only be constant in strength whilst the proportion of electromotive force (or difference of potential) to resistance remains unchanged. Generally if V be the potential difference in volts at the ends of a conductor whose resistance is R ohms, and if A be the current in amperes passing through it, then $A = \frac{V}{R}$. Ohm's law shows that the strength of current passing through any circuit is inversely proportional to the resistance if a constant potential difference is maintained at the ends of that circuit.

A unit current in amperes is that strength of current which deposits ·001118 gramme or ·01725 grain of silver per second from an aqueous solution containing fifteen to thirty per cent. of argentic nitrate, or ·0003296 gramme or ·00508 grain of copper from a solution of blue vitriol per second.

Measurement of Strength of Current.—Strength of current is usually ascertained, either (1) by means of its chemical effect, as in a voltameter (Fig. 7), by finding how much hydrogen, silver, or copper is set free by electrolysis in a given time (*see* p. 105); (2) by its magnetic influence, as in a galvanometer (Figs. 8 and 9), by observing the amount of deflection of a magnetised needle or to what extent a coil conveying the current is deflected by a permanent magnet; or, as in an electro-dynamometer (Fig. 10), by noting through what angle of deflection a coil of wire conveying one portion of the current is moved by the influence of a second coil conveying another portion; or (3) by the rate of production of its heat of conduction resistance. This latter method, as well as that in which

c 2

an electro-dynamometer is employed, is used in measuring the strength of alternating currents. It is only when the current is sufficiently strong to quickly raise the temperature of the liquid of a calorimeter several degrees that method "3" can be employed. For particulars respecting the methods of measurement consult the works already referred to on p. 17.

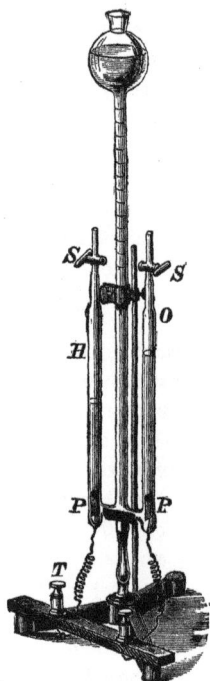

FIG 7.—Hofmann's Voltameter.   FIG. 8.—Torsion Galvanometer.

Density of Current.—By this is meant the strength of current passing at a given moment through a given cross-sectional area of a conductor, or into or out of a given sized surface of an electrode in an electrolyte.

Circuit.—The entire path of the current is termed the circuit. The amounts of steady current passing at the same moment through all cross-sections of a circuit are equal, and are independent of the composition, shape, size, or number of pieces of the conductor. In many cases it is only after the first instant of flow that the current is of the same strength throughout the circuit; this is due to self-inductive action, which precedes conduction.

A SHUNT is a divided circuit, one portion of the current being shunted or turned aside from the main circuit into a side one (Fig. 11). When a portion of a main current is shunted, the

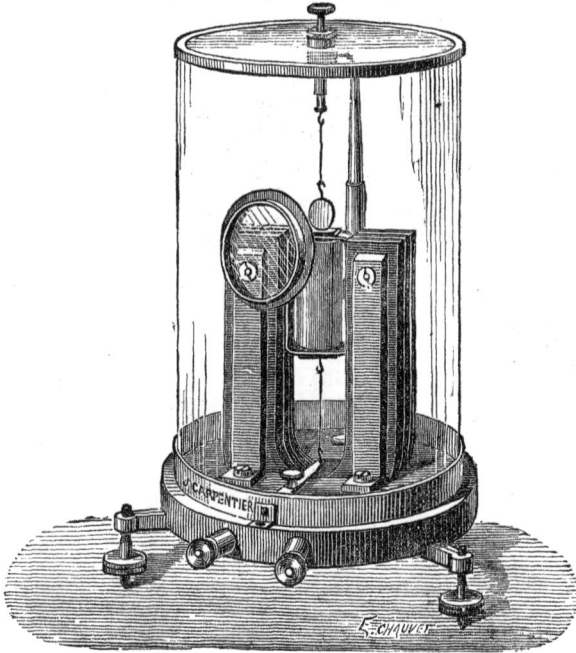

FIG. 9.—D'Arsonval's Galvanometer.

strength of the main current parallel with the shunt is decreased, whilst that before and behind the shunt is increased, because the total resistance in the circuit is diminished by adding the shunt conductor, this addition being equivalent to increasing the thickness of that portion of the main conductor. The sum of the strengths of current in the divided part of the circuit equals the strength of the main current; and the

relative strengths in the divided part are inversely as the
relative resistances of the two parallel conductors.

The conductors in a circuit may be arranged either in single
series, *i.e*, one after another, thus — — — —; in parallel, *i.e.*,
side by side ☰; or in combined series and parallel, thus = = =
or ☰ ☰.

Practically in all cases where a current from a generator of
very small internal resistance is employed, the currents in
parallel circuits supplied by it are largely independent of each

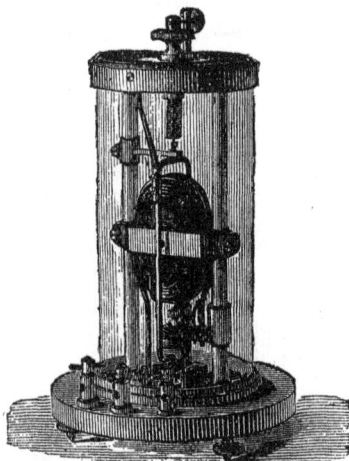

FIG. 10.—Siemens's Electro-Dynamometer.

other. The strength of current in any one of the circuits
depends upon the resistance of the particular circuit and the
potential difference at the poles of the generator.

By "short-circuiting" is meant the cutting-out of a portion
or the whole of the external resistance by means of a conductor
of less or very small resistance, so that the circuit is greatly
diminished in resistance and usually shortened.

A SWITCH is a contrivance, either worked by hand or
automatically, for diverting the current into a separate course,
generally when the current in the usual circuit is from some
cause altered or interrupted. An automatic one is generally
operated by the magnetic influence of the current itself, which,

when the latter becomes too strong or too weak, liberates a mechanism and makes a fresh contact.

A CUT-OUT is a similar contrivance for breaking the circuit, either when the current becomes too weak, too strong, or is reversed in direction; it is either automatic or not. In an automatic one, an electro-magnet breaks the circuit. An automatic cut-out is sometimes formed of a fusible wire or wires, usually of lead about two or three inches long, inserted in the main circuit on either side of the generator, so that when from any cause the current becomes too strong the wire melts and breaks the circuit.

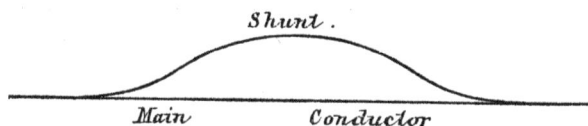

FIG. 11.—Shunt Conductor.

Electrical Energy.—This includes electromotive force and strength of current. The practical unit of electric energy is termed a WATT, and is the work done per second by one ampere passing between two points, between which the difference of electric potential is one volt. It is $= 10^7$ ergs ; also ·7375 foot-pound, or $\frac{1}{746}$th of a horse-power (or $\frac{1}{736}$ French horse-power), one horse-power being $= 550$ foot-pounds per second. To find the amount of electric energy of a current in watts, multiply the number of its amperes by that of its volts. The number of watts or volt-amperes being expended in any circuit may be measured directly by means of a wattmeter.

Conduction and Insulation.—The conducting and insulating powers, or the capacities of substances to permit and to hinder electric flow, differ in degree in every different substance ; all substances permit and all hinder in different degrees, and the difference of degree of these properties in extreme cases is enormous. The best insulators (called also dielectrics) are ebonite, shellac, india-rubber, gutta-percha, resins, wax, asphalte, gases, glass, &c., and the best conductors are the metals, especially the ductile ones, copper, silver, and gold. Solutions of acids, alkalies, and metallic salts, occupy an intermediate posi-

tion. If we represent the resistance of silver as being equal
to 1, that of a piece of gutta-percha of the same dimensions
has been estimated to be about 850 million, million, millions.
A current which rapidly alternates in direction, passes chiefly
through the surface of a conductor.

Leakage of Electricity.—This is often a serious matter,
and may be divided into that of charge and that of current.
Electric charge may leak and pass away either over the surface of a body or through its mass ; it may also be dissipated
by convection, by the dust of the air being attracted,
charged, and then repelled.   Surface leakage of charge is
usually due to dust or moisture, and varies in amount according to the ordinary laws of conduction resistance, *i.e.*, the
greater the width and the shorter the length of the conducting film, the greater the leakage.   That through the
mass of an insulating substance diminishes gradually until
the substance becomes charged to some depth, and then
remains constant ; it increases by rise of temperature.   Long
rods of non-conducting substance insulate more perfectly than
short ones, both on their surface and in their mass.   Leakage of currents of small electromotive force may be due to
various causes, usually to partial short-circuiting caused by
accidental metallic contacts, spilling of acids or of saline
liquids, by water soaking into wood, &c.

Electric Conduction Resistance.—All substances resist more
or less the passage of an electric current.   Electric conduction
resistance is that quality of a substance which prevents more
than a certain amount of current passing through it in a given
time when impelled by a given electromotive force, or whilst
the two ends of the substance are maintained at a given difference of electric potential.   If two conducting bodies at different
potentials are suddenly united by a conductor it takes a sensible
amount of time for the potentials to become equalised and for
the electricity to pass through the conductor ; this delay differs
in amount with different substances, and indicates a hindrance
in the conductor.

The resistance of a conductor is equal to the time required
for a unit quantity of electricity (*i.e.*, a coulomb) to pass through

it whilst its two ends are maintained at unit difference of potential, *i.e.*, at one volt. The ordinary practical unit of resistance is termed an OHM. A British Association ohm is equal to 10^9 absolute C.-G.-S. electro-magnetic units of resistance, or to that of a column of mercury 1·0486 metre long and 1·0 square millimetre section at 0°C. The legal ohm, as settled by the International Congress held in Paris in 1884, is slightly different from this, and is 1·060 metre long ; one legal ohm = 1·0112 British Association unit, the previously used practical unit of resistance. At 41·5°C. the B. A. unit coil has a resistance equal to that of the legal ohm at 0°C. A microhm is a millionth of an ohm. A megohm is a million ohms.

Liquids exhibit extremely different degrees of electric conduction resistance; whilst some easily transmit the current from a single feeble voltaic cell, others completely resist the passage of a current from ten thousand such elements in single series. Amongst the least resisting ones are dilute acids, solutions of salts of copper, silver, gold, lead, zinc, potassium, sodium, and of the metals generally. The most resisting ones, or the best insulators, are petroleum, benzine, oils, bisulphide of carbon, ether, &c. Perfectly pure water is a very bad conductor.

The conduction resistance of liquids in comparison with that of a metal is enormous, and is shown by the following table :—

Copper at 0°C.	1
Nitric acid at 13°C.	976,000
Sulphuric acid diluted to $\frac{1}{71}$th at 20°C.	1,032,000
Saturated solution of common salt, at 13·5°C.	2,903,538
,, ,, zinc vitriol	15,861,267
,, ,, blue ,, at 9°C.	16,885,520
Distilled water at 15°C.	6,754,208,000

The resistance of a wire is directly proportional to its length, and inversely to its sectional area, or weight per unit of length ; the sectional areas of wires are proportional to the squares of their diameters (*see* Appendix). No. 50 copper wire, B.W.G., weighing $\frac{1}{20}$ gramme per metre, has a resistance of about 3 ohms per metre. Conductivity is the reciprocal of resistance.

Table of Conducting Powers of Metals (Matthiessen).

	Conducting powers.		Conducting powers.
Silver...............	100·0	Tin..................	12 4
Copper	99·9	Thallium	9·2
Gold	77·9	Lead	8·3
Zinc	29·0	Arsenic	4·8
Cadmium	23·7	Antimony	4·6
Platinum	18·0	Mercury	1 6
Cobalt	17·2	Bismuth	1·2
Iron	16·8	Graphite	·069
Nickel	13·1	Gas coke	·038

Conductivity of Copper (Matthiessen).

	Conductivity.		Degrees Centigrade.
Chemically pure copper.....................	100·0	at	...
Lake Superior, native, not fused.........	98·8	,,	15·5
,,　　　,,　　　,,　 fused	92·6	,,	15·
Burra-Burra	88 7	,,	14·
Best selected	81·3	,,	14·2
Bright copper wire.........................	72·2	,,	15·7
Tough copper	71·0	,,	17·3
Demidoff	59·3	,,	12·7
Rio Tinto	14·2	,,	14·8

Table of Conduction Resistances of Metals (Matthiessen).

—	Relative Resistances.	Resistance in legal microhms of a centimetre cube at 0°C.	Resistance in legal ohms of a wire 1 metre long, weighing 1 gramme.
Silver, annealed	1·000	1·504	·1527
Copper ,, ...	1·063	1·598	·1424
Silver, hard drawn	1·086	1·634	·1662
Copper ,, ,, ...	1·086	1·634	·1453
Gold, annealed	1·369	2·058	·4035
,, hard drawn...............	1·393	2·094	·4104
Aluminium, annealed	1·935	2·912	·0749
Zinc, pressed.....................	3·741	5·626	·4023
Platinum, annealed	6·022	9·057	1·938
Iron ,, ...	6·460	9·716	·757
Alloy, gold 2 pts., silver 1 pt., hard or annealed......	7·228	10·87	1·650
Nickel, annealed	8·285	12·47	...
Tin, pressed	8·784	13·21	·9632
Lead ,, ...	13·05	19·63	2·232
German silver, hard or annealed	13·92	20·93	1·830
Alloy, silver 2 pts., platinum 1 pt., hard or annealed ...	16·21	24·39	2·924
Antimony, pressed	23·60	35·50	2·384
Mercury..........................	62·73	94·32	12·91
Bismuth..........................	87·23	131·20	12·88

Relative Conductivity of Alloys.

The following is from a table given by L. Weiller :—

Pure silver	100·
Silicon bronze telegraph wire	98·
Copper and silver alloy of 50 per cent.	86·65
Silicon copper with 4 per cent. of silicon	75·
,, ,, 12 ,, ,,	54·7
Tin containing 12 per cent. of sodium	46·9
Silicon bronze telephone wire	35·0
Plumbic copper, with 10 per cent. lead	30·
Phosphor bronze telephone wire	29·
Silicon brass, with 25 per cent. of zinc	26·49
Brass, with 25 per cent. of zinc	21·15
Phosphide of tin	17·7
Gold and silver alloy 50 per cent.	16·12
Antimonic copper	12·7
Aluminium bronze 10 per cent.	12·6
Siemens's steel	12·0
Cadmium amalgam, with 15 per cent. of cadmium	10·2
Mercurial bronze	10·14
Arsenical copper, with 10 per cent. of arsenic	9·1
Bronze, with 20 per cent. of tin	8·4
Phosphor bronze, with 10 per cent. of tin	6·5
Phosphide of copper, with 9 per cent. of phosphorus	4·9
Antimony	3·88

An alloy called "platinoid," consisting of nickel silver with about 1 or 2 per cent. of tungsten, has been found by Mr. Bottomley to possess a resistance of about 60 per cent. higher than that of German silver.

Relative Conductivities of Alloys of Copper (Matthiessen).

Pure Copper=100.

	Conductivity.		Temperature in Centigrade degrees.
·5 per cent. of carbon	77·87	at	18·3
·18 ,, ,, sulphur	92·08	,,	19·4
·13 ,, ,, phosphorus	70·34	,,	20·0
·95 ,, ,, ,,	24·16	,,	22·1
2·5 ,, ,, ,,	7·52	,,	17·5
With traces of arsenic	60·08	,,	19·7
2·8 per cent. ,,	13·66	,,	19·3
5·4 ,, ,,	6·42	,,	16·8
With traces of zinc	88·41	..	19·0
1·6 per cent. of zinc	79·37	··	16·8
3·2 ,, ,, ,,	59·23	,.	10·3
·48 ,, ,, iron	35·92	,,	11·2
1·66 ,, ,, ,,	28·01	,,	13·1
1·33 ,, ,, tin	50·44	,,	16·8
2·52 ,, ,, ,,	33·93	,,	17·1
4·90 ,, ,, ,,	20·24	,,	14·4
1·22 ,, ,, silver	90·34	,,	20·7

Relative Conductivities of Alloys of Copper (Matthiessen), continued.

Pure Copper = 100.

			Conductivity.		Temperature in Centigrade degrees.
2·45	per cent, of silver		82·52	at	19·7
3·5	,,	,, gold	67·94	,,	18·1
10·0	,,	,, aluminium	12·68	,,	14·C
·31	,,	,, antimony ⎱	64·5	,,	12·0
·29	,,	,, lead ⎰			

Resistance of Pure Copper Wire at 0°C.

Diameter in millimeters.	Metres per ohm.	Diameter in millimetres.	Metres per ohm.
·5	12·305	9·0	3956·
1·0	48·87	10·0	4878·
1·5	109·75	11·0	5854·
2·0	195·15	12·0	7025·
2·5	308·6	13·0	8293·
3·0	439·1	14·0	9680·
3·5	605·	15·0	10975·
4·0	777·	16·0	12032·
4·5	989·	17·0	14450·
5·0	1231·	18·0	15824·
6·0	1756·	19·0	17561·
7·0	2420·	20·0	19515·
8·0	3008·		

Electric Resistance of Metals.

(Benoit, *Comptes Rendus*, Paris, 1873, Vol. LXXVI., p. 342.)
(*Journal* of the Society of Telegraph-Engineers, Vol. I., p. 443.)

1 metre long × 1· sq. mm. section at 0°C.

	Ohms.	Conductivity.
Silver, pure, annealed	·0154	100·
Copper ,, ,,	·0171	90·
Silver ($\frac{75}{100}$) ,,	·0193	80·
Gold, pure ,,	·0217	71·
Aluminium, pure, annealed	·0309	49·7
Magnesium, hammered	·0423	36·4
Zinc, pure, annealed at 350°C	·0565	27·5
,, ,, hammered	·0594	25·9
Cadmium, pure, hammered	·0685	22·5
Brass, annealed	·0691	22·3
Steel ,,	·1099	14·0
Tin, pure	·1161	13·3
Aluminium bronze, annealed	·1189	13·0
Iron, annealed	·1216	12·7
Palladium, annealed	·1384	11·1
Platinum ,,	·1575	9·77
Thallium	·1831	8·41
Lead, pure	·1985	7·76
Maillechort, annealed	·2654	5·80
Mercury, pure	·9564	1·61

For a given length and weight, aluminium has the least resistance of all metals, but for a given length and diameter annealed silver has the least. The order of conductivity of metals for electricity is nearly the same as that for heat; but with less conducting substances the conductivity for electricity diminishes very much more rapidly than that for heat; conse-quently the insulation of electricity is vastly less difficult than that of heat.

The resistance of atmospheric air and gases is usually much greater than that of liquids. Liquids conduct, according to Ohm's law, the same as solids (*Journal* of Chemical Society, 2nd Series, Vol. X., p. 208). The resistance of dilute hydro-chloric acid, and of a solution of zinc sulphate, is said to be increased by a pressure of five hundred atmospheres (*Nature*, Vol. XXXIII., 1886, p. 356). According to Dr. Overbeck, the conduction resistance of ether is 102 times larger than that of water, and that of bisulphide of carbon is still greater.

Fig. 12.—Wheatstone's Bridge.

The amount of resistance of a wire, A, may be conveniently measured by dividing a current from a very small Daniell's cell, so that one portion passes through A and on through one wire B of a differential galvanometer, and the other portion through another wire C of known amount of resistance, such as a British Association standard unit of electrical resistance, and on through the other wire D of the galvanometer in the opposite direction to that through B, and then altering the length of A until the needles of the instrument stay at zero; the resistance of A and C are then equal, provided the resist-ances and the lengths of the two wire coils of the galvanometer are alike. For more accurate measurements a Wheatstone's bridge should be employed (Fig. 12). Various other methods are described in books devoted to the subject (*see* foot note, p. 17).

Resistance of Solutions of Sulphuric Acid (H_2SO_4)
(Matthiessen).

Specific Gravity of Liquid.	Percentage of H_2SO_4 by weight.	Resistance.	Temperature in Centigrade degrees.
1·003	·5	16·01	16·1
1·018	2·2	5·47	15·2
1·053	7·9	1·88	13·7
1·080	12·0	1·37	12·8
1·147	20·8	·96	13·6
1·190	26·4	·87	13·0
1·215	29·6	·83	12·3
1·225	30·9	·86	13·6
1·252	34·3	·87	13·5
1·277	37·3	·93	—
1·348	45·4	·97	17·9
1·393	50·5	1·09	14·5
1·493	60·6	1·55	13·8
1·638	73·7	2·79	14·3
1·726	81·2	4·34	16·3
1·827	92·7	5·32	14·3

Resistance of Solutions of Cupric Sulphate ($CuSO_4 + 5H_2O$) at 14°C. (Wiedemann).

Grammes per litre of water.	Ohms.	Grammes per litre of water.	Ohms.
31·17	124·3	124·68	45·7
62·34	70·2	155 85	39·9
77·92	59·9	187·02	36·3
93·51	54·8		

The resistance of solutions of sulphate of copper and sulphate of zinc increases regularly by dilution of the liquids from their points of saturation, but that of a solution of chloride of sodium has the least resistance when the liquid contains about 24 per cent of the salt.

Specific Resistance of Solutions of Sulphate of Copper at 10°C. (Ewing and MacGregor).

Specific Gravity.	Resistance.	Specific Gravity.	Resistance.
1·0167	164·4	1·1386	35·0
1·0216	134·8	1·1432	34·1
1·0318	98·7	1·1679	31·7
1·0622	59·0	1·1823	30·6
1·0858	47·3	1·2051 (saturated)	29·3
1·1174	38·1		

The measurement of resistance of an electrolyte is much less easy than that of a wire, in consequence of the varying degree of polarisation and counter electromotive force produced at the electrodes by the passage of the current. It may, however, be approximately effected in a somewhat similar manner (see p. 29) by making two measurements with a very feeble current after the polarisation has become sufficiently steady, one when the electrodes are near together, and the other when they are farther asunder; using in each case a narrow trough filled with the liquid, with two immersed porous cells containing electrodes as large as the transverse section of the electrolyte, and usually of the same metal as that of the salt of the liquid. The difference of resistance in the two measurements is that of the difference in length of the liquid in the two cases. The liquids in the porous cells should be rapidly stirred in order to diminish or prevent polarisation whilst making the measurements.

Influence of Temperature on Resistance.—Usually, by rise of temperature, the conduction resistance of metals is increased, and that of electrolytes and gases is decreased ; with carbon and mercury it is decreased. That of copper increases about 1 per cent. by a rise of 2·57 centigrade degrees. According to E. Weston, an alloy, composed of 65 to 70 parts of copper, 25 to 30 of ferro-manganese, and $2\frac{1}{2}$ to 10 of nickel, has its resistance *lowered* by rise of temperature (*The Electrician*, Vol. XXI., August 10th, 1888, p. 448). For the resistance of metals at high temperatures, consult Benoit's research in the *Comptes Rendus* of the Paris Academy of Sciences, January to June, 1873, p. 342. According to Matthiessen and Benoit, the coefficient of electric conduction resistance is about $\frac{1}{273}$rd of the absolute resistance for each centigrade degree rise of temperature between 0°C. and 100°C.

According to Clausius, the resistance of copper is proportional to the absolute temperature ; therefore, at the absolute zero of heat, or − 273°C., the conducting power of that metal would be infinite. Cailletet and Bouty have experimented in this direction down to − 123°C. with copper and other metals, and conclude as probable, that with certain ductile metals the resistance below − 200°C. would be very small. Wroblewski has also made similar experiments, and found that whilst the resistance of copper at + 100°C. was 5·174 Siemens's units, at − 200°C. it was only ·414.

Relative Resistances of Pure Copper at Different Temperatures.

Temp. in °C.	Resist.	Temp. in °C.	Resist.	Temp. in °C.	Resist.
0	1·0000	10	1·0382	21	1·0816
1	1·0038	11	1·0420	22	1·0855
2	1·0076	12	1·0460	23	1·0895
3	1·0114	13	1·0500	24	1·0936
4	1·0152	14	1·0541	25	1·0976
5	1·0190	15	1·0577	26	1·1016
6	1·0228	16	1·0617	27	1·1057
7	1·0266	17	1·0656	28	1·1197
8	1·0305	18	1·0696	29	1·1238
9	1·0344	19	1·0736	30	1·1278
		20	1·0774		

The following table embodies some results obtained by the late
Dr. Matthiessen on the influence of temperature on resistance :

Percentage Variation per 1°C. Degree at about 20°C.

Alloy, platinum 2 pts., silver 1 pt., hard or annealed, about ·037
German silver, hard or annealed .... ,, ·044
Alloy, gold 2 pts., silver 1 pt., hard or annealed ... ,, ·065
Mercury.. ,, ·072
Bismuth, pressed .... ,, ·354
Gold, annealed .... ,, ·365
Zinc, pressed .... ,, ·365
Tin   ,, ... ,, ·365
Silver, annealed .... ,, ·377
Lead, pressed .... ,, ·387
Copper, annealed... ,, ·388
Antimony .... ,, ·389
Iron... ,, ·500

The alloy termed platinoid (*see* p. 27) has a percentage variation of only about ·021 per 1 centigrade degree, or about half
that of German silver, and as its conduction resistance is higher
by 60 per cent. than that of that alloy, it may prove suitable
for resistance coils if it does not change by lapse of time.

*Resistance in Ohms of a Cubic Centimetre of Various Solutions
of Cupric Sulphate at Different Temperatures (Centigrade).*
(Computed by Fleeming Jenkin.)

Percentage of Dissolved Salt.	Specific Gravity at 18°C.	At 14°C.	16°.	18°.	20°.	24°.	28°.	30°.
8	1·0516	45·7	43·7	41·9	40·2	37·1	34·2	32·9
12	1·0785	36·3	34·9	33·5	32·2	29·9	27·9	27·0
16	1·1063	31·2	30·0	28·9	27·9	26·1	24·6	24·0
20	1·1354	28·5	27·5	26·5	25·6	24·1	22·7	22·2
24	1·1659	26·9	25·9	24·8	23·9	22·2	20·7	20·0
28	1·1980	24·7	23·4	22·1	21·0	18·8	16·9	16·0

Resistance in Ohms of a Cubic Centimetre of Solutions of Sulphuric Acid at Different Temperatures (Centigrade). (Computed by Fleeming Jenkin.)

Specific Gravity of Liquid at 15° C.	Containing per cent. of $H_2 SO_4$.	At 0°.	4°.	8°.	12°.	16°.	20°.	24°.	28°.
1·10	15·	1·37	1·17	1·04	·925	·845	·786	·737	·709
1·20	27·	1·33	1·11	·926	·792	·666	·567	·486	·411
1·25	33·	1·31	1·09	·896	·743	·624	·509	·434	·358
1·30	40·	1·36	1·13	·94	·79	·662	·561	·472	·394
1·40	50·	1·69	1·47	1·30	1·16	1·05	·964	·896	·839
1·50	60·	2·74	2·41	2·13	1·89	1·72	1·61	1·52	1·43
1·60	68·	4·82	4·16	3·62	3·11	2·75	2·46	2·21	2·02
1·70	77·	9·41	7·67	6·25	5·12	4·23	3·57	3·07	2·71

According to the computation of the same writer, the resistance of one cubic centimetre of a solution of zinc sulphate, containing 96 grammes of the salt in 100 c.c. of solution, was = 22·7 at 10°C., and decreased regularly to 15·6 at 24°C. And when the same solution was diluted with an equal bulk of water, the resistance at 14·0°C. was 21·1 ohms, and decreased regularly to 17·3 ohms at 24°C. For further information respecting the conduction resistance of a large variety of electrolytes, see tables compiled from the researches of F. and W. Kohlrausch, Grotian, Long, and others, in Landolt and Bernstein's " Physikalische-Chemische Tabellen," 1886, pp. 100-107.

Conduction Resistance of Minerals, &c.—In consequence of the great variations of density and of physical structure in different specimens of the same compound, the usual difficulty of obtaining suitable specimens for measurement, and other circumstances, minerals and solid chemical compounds can only at present be crudely classified into inferior conductors and relatively non-conductors. To the former group belong magnetic iron ore, tinstone, peroxide of lead, arsenical silver, red silver, galena, arsenical cobalt, copper glance, cubical iron pyrites, arsenical iron pyrites, magnetic pyrites, peroxide of manganese, sulphide of bismuth; the sulphides of mercury, copper, nickel and cobalt; mispickel, tin pyrites, subsulphide of copper, mag-

D

netic iron ore, graphitic tellurium, oxide of zinc, also nitro-
cyanide of titanium and titanic iron ore. In the latter group
are included zinc blende, sulphide of molybdenum, crystallised
stibnite, compact variety of crystallised cinnabar, orpiment,
bouranite, manganese blende, proustite, pyrargite, silver-glance,
horn-silver, calamine, crystallised chrome ore, crystallised black
carbonate of iron, crystallised tungstate of iron, rutile, braunite,
crystallised specular iron ore, iserine, crystallised tin ore, suboxide and protoxide of copper, sulphide of silver, oxychloride of
copper, &c. Whilst cupreous sulphide is a very bad conductor,
cupric sulphide is a relatively good one. The latter fact
is important in certain practical electrolytic processes (*see*
pp. 228-237).

Internal and External Resistance.—The total resistance in
an electrolytic circuit is usually classed into internal, or that in
the electric generator, and external, or that in the electrolysis
vessel and the remainder of the circuit.

"Transfer Resistance."—By this term is meant a retardation
at the surfaces of electrodes in electrolytes, different from that due
to polarisation or other counter electromotive force (*see* p. 95).
Evidence respecting it may be found in the following published
researches by the author: " On 'Transfer-resistance' in Electrolytic and Voltaic Cells" (Abstract, *Proc.* Roy. Soc., Vol. XXXVIII.
March, 1885, p. 209). "On 'Resistance' at the Surfaces of
Electrodes in Electrolytic Cells" (*Proc.* Birm. Phil. Soc.,
Vol. V., p. 45; *Phil. Mag.*, Vol. XXI., 1886, p. 249). "Relation
of 'Transfer-resistance' to the Molecular Weight and Chemical
Composition of Electrolytes" (Abstract, *Proc.* Roy. Soc., 1886,
p. 380; full paper, *Proc.* Birm. Phil. Soc., Vol. V., 1887,
p. 426). "Evidence respecting the Reality of 'Transfer-resistance' in Electrolytic Cells" (*Proc.* Birm. Phil. Soc., Vol. V.,
p. 26; *Phil. Mag.*, 1886, Vol. XXI., p. 130). "Relation of
Surface-resistance at Electrodes to Various Electrical Phenomena" (*Proc.* Birm. Phil. Soc., Vol. V., p. 36; *Phil. Mag.*,
1886, Vol. XXI., p. 45). "Influence of External Resistance on
Internal Resistance in Voltaic Cells" (*Proc.* Birm. Phil. Soc.,
Vol. IV., p. 417; *The Electrician*, Vol. XV., p. 279). W. Peddie
has also investigated this phenomenon. He states "conclusively

that a transition resistance exists," and that with platinum plates in dilute sulphuric acid it " is due to condensed films of gas " upon the surface of the electrode, and is not removed by hard rubbing. He also found that " the resistance is inversely proportional to the area of the plates," and that " the order of the specific resistance is the same as that of ordinary dielectrics " (" On Transition-resistance at the Surface of Platinum Electrodes, and the Action of Condensed Gaseous Films," *Proc.* Roy. Soc. Edinburgh, 1886-1887, No. 124). I have verified the fact that the gas absorbed by a platinum cathode in water is not wholly removed by hard rubbing, but is expelled by heating to redness. It is not improbable that this gas offers resistance to the passage of the current, independently of that due to its well-known counter electromotive force. The physical actions which occur at the surfaces of electrodes in electrolytes are more complex than is sometimes assumed. J. Monckman, *Proc.* Roy. Soc., May 31st, 1888, pp. 223-226, has shown that platinum and iron, by absorbing hydrogen when employed as cathodes, increase in conduction-resistance. These circumstances are sufficient to account for the " resistance " I observed.

SECTION B.

CHIEF THERMAL PHENOMENA OF THE SEPARATION OF METALS BY MEANS OF DYNAMO ELECTRIC CURRENTS.

The chief thermal phenomena taking place in dynamo-electric machines are :—1. Heat evolved by conduction resistance in the conducting wires surrounding the armature and field-magnets. 2. Heat generated by Foucault or eddy induction currents in the metallic portions of the magnet and armature. 3. Heat produced by friction of the axles of the armatures in their bearings.

The chief source of thermal changes in the electrolysis vessel appear to be :—1. Heat evolved or absorbed by chemical and electro-chemical changes at each of the electrodes. 2. Heat produced by conduction resistance of the electrolyte. 3. By conduction resistance of gaseous, liquid or solid films, formed upon the anode or cathode. 4. By resistance due to polarisation at each of the electrodes. 5. By "transfer resistance" at those electrodes. 6. By chemical action in any sediment at the bottom of the vessel.

Heat of Conduction Resistance.—Whenever a current passes through a resistance it evolves heat, and as the best of conducting substances offers resistance, the passage of a current is always attended by evolution of heat. The amount of heat thus produced by a current in a conductor in a given time is directly proportional to the product of the square of the strength of the current into the resistance. This is known as Joule's law. If the composition and diameter of a wire are uniform throughout its length, with the same strength of current passing through its entire length, and uniform cooling influences, the increase of temperature is the same in all parts of the wire.

The electro-thermal unit of conduction resistance, termed a *joule*, is the amount of heat produced in one second by a current of one ampere flowing through a resistance of one ohm, and is the quantity necessary to raise the temperature of ·239 gramme of water one centigrade degree. One joule is equal to ·7375 foot-pound, or to 1 watt of power exerted during one second; it is only ·24 of an ordinary heat unit or centigrade-gramme calorie (*see* below).

Mechanical Equivalent of Heat.—Joule's mechanical equivalent of heat is the amount of energy required to raise 772·55 lbs. one foot high, and is equal to the quantity of heat necessary to increase the temperature of one pound of water at 60° Fahrenheit one Fahrenheit degree. The quantity of heat required to raise the temperature of one gramme of water at 0°C. one centigrade degree is equal to the energy necessary to lift 423·55 grammes or ·42355 kilogramme weight one metre high, or to that of 3·0636 foot-pounds. (One kilogrammetre = 7·233 foot-pounds or 9·807 joules.)

Thermal Units.—The unit of quantity of heat is termed a *calorie*, and is that amount which will raise the temperature of a unit mass of water, at some assumed standard temperature, through one thermometric degree. The unit of mass employed is usually either one gramme, pound or kilogramme. If it is a gramme raised one centigrade degree it is called a centigrade gramme calorie; and if a kilogramme, it is a centigrade kilogramme calorie. One gramme of water at about 18°C. raised one centigrade degree is a commonly used unit, and is equal to ·0098 joule or electro-thermal unit (*see* Appendix).*

Heat of Chemical Union and Decomposition.—In every chemical action, and in every voltaic and electrolytic one, there is a transformation of energy, an evolution or absorption of heat, and a change of physical condition of the substances. In every voltaic and electrolytic action substances either unite

---

* The centigrade-gramme calorie is now called a THERM, and the term JOULE is used to indicate the work done by one watt in one second; 4·2 joules = 1 therm (O. J. Lodge, *The Electrician*, Vol. XXI., September 28th, 1838, p. 661).

together chemically to form compounds, or compounds are chemically decomposed. Chemical combination very usually evolves, and decomposition nearly always absorbs heat. Whatever amount of heat two substances evolve when combining, that same amount do they absorb when separating. The amounts of heat liberated by the elementary substances when chemically combining vary in substantially the same order as the degrees of chemical potential of the substances (*see* p. 48), and as the position of the substances in the volta-tension series (*see* p. 50).

The amount of heat or thermal energy necessary to decompose a given compound is equal to that produced by the separated substances when uniting together to form that compound in the same physical state. This agrees with the principle of conservation of energy, which affirms that in any substance or system of substances the total amount of heat or energy in it is a fixed quantity, and is unaffected by any intervening changes which occur, provided the substance or system is in exactly the same physical and chemical condition after the changes that it was in before they occurred.

Thermal Symbols and Formulæ.—The thermal changes which take place in chemical and electro-chemical actions are represented by symbols and formulæ, very much like those used to represent ordinary chemical changes. Thermo-chemical formulæ are generally enclosed in a small bracket; a comma, or sometimes a colon, being placed between the two formulæ of the substances which react upon each other. The plus or negative signs prefixed or affixed to the numbers representing heat of formation, of solution, or of decomposition, indicate whether the action evolves or absorbs heat; if no sign is given plus is meant. For instance $(H^2, O) = +68360$ means that when two atomic weights of hydrogen in grammes unite with one atomic weight of oxygen in grammes at a temperature of $18°$ to $20°C.$, and form 18 grammes of water at that temperature, sufficient heat is set free to raise the temperature of 68360 grammes of water at about $18°C.$ one centigrade degree. Substances whose symbols are not separated by a comma or colon have already united together chemically and evolved (or absorbed) heat. The symbol Aq indicates an unlimited amount of water.

The following tables give the relative quantities of heat, in centigrade-gramme units, evolved by the chemical union of various substances with each other, in the proportions of their equivalent weights in grammes as represented by the formulæ given. The numbers are those given by Thomsen in his "Thermo-Chemical Memoirs," and are results of his laborious investigations :—

Heat of Chemical Union.

Chemical Reaction.	Centigrade-Gramme Calories.	Heat of Solution in Abundance of Water.
(H^2, O)	+68360	
(H^2, S)	4740	4560
(H^2, S, Aq)	9300	
(H, Cl)	22000	17315
(H, Cl, Aq)	39315	
(H, Br)	8440	19940
(H, I)	− 6040	19210
(H^3, N)	11890	8340
(C, O)	29000	
(C, O^2)	96960	
(S, O^2)	71080	
(S, O^3)	103240	
(S, O^3, H^2O)	124560	
(S, O^3, Aq)	142410	
(P^2, O^5)	369000	
(As, Cl^3)	71390	17580
(As^2, O^3)	154670	− 7550
(As^2, O^5)	219380	6000
$(As^2, O^5, 3H^2O)$	226180	
(Sb, Cl^3)	91390	
(Sb, Cl^3, Aq)		7730 Entirely decomposed.
$(Sb^2, O^3, 3H^2O)$	167420	
$(Sb^2, O^5, 3H^2O)$	228780	
(Bi, Cl^3)	90630	
(Bi, O, Cl, H^2O)	88180	
$(Bi^2, O^3, 3H^2O)$	137740	
(Au, Cl^3)	22820	4450
$(Au, Cl^3, 2H^2O)$	28960	− 1690
(Ag^2, O)	5900	
(Ag^2, Cl^2)	58760	
(Ag^2, O^2, SO^2)	96200	− 4480
(Hg, O)	30670	
(Hg^2, O)	42200	
(Hg, Cl^2)	63160	− 3300
(Hg^2, Cl^2)	82550	
(Cu, O)	37160	
(Cu^2, O)	40810	

Heat of Chemical Union (continued).

Chemical Reaction.	Centigrade-Gramme Calories.	Heat of Solution in Abundance of Water.
(Cu^2, O, H^2O)	37520	
(Cu, Cl^2)	51630	11080
(Cu, Br^2)	32586	8250
(Cu, O^2, SO^2)	111490	15800
$(Cu, O^2, SO^2, 5H^2O)$	130040	− 2750
$(Cu, O^2, N^2O^4, 6\ H^2O)$	96950	−10710
(Ni, Cl^2)	74530	19170
(Ni, O, H^2O)	60840	
$(Ni, O^2, SO^2, 7H^2O)$	162530	− 4250
(Co, Cl^2)	76480	18340
(Co, O, H^2O)	63400	
$(Co, O^2, SO^2, 7H^2O)$	162970	− 3570
(Fe, Cl^2)	82050	17900
$(Fe, Cl^2, 4H^2O)$	97200	2750
(Fe^2, Cl^6)	192080	63360
(Fe, O, H^2O)	68280	
$(Fe, O^2, SO^2, 7H^2O)$	169040	− 4510
(Mn, Cl^2)	111990	16010
(Mn, O, H^2O)	94770	
(Mn, O^2, H^2O)	116330	
(Mn, O^2, SO^2)	178790	13790
(Al^2, Cl^6)	321960	
(Tl^2, O^2, SO^2)	149900	− 8280
(Pb, Cl^2)	82770	− 6800
(Pb, O)	50300	
$(Pb^2, O^3, 3H^2O)$	250320	
(Pb, O^2, SO^2)	145130	
(Sn, Cl^2)	80790	350
$(Sn, Cl^2, 2H^2O)$	86560	− 5370
(Sn, O, H^2O)	68090	
(Sn, O^2, H^2O)	133500	
(Cd, Cl^2)	93240	3010
(Cd, O, H^2O)	65680	
(Cd, O^2, SO^2)	150470	10740
(Zn, Cl^2)	97210	15630
(Zn, O, H^2O)	82680	
(Zn, O^2, SO^2)	158990	18430
(Mg, Cl^2)	151010	35920
$(Mg, Cl^2, 6H^2O)$	183980	2950
(Mg, O^2, SO^2)	231230	20280
(Sr, Cl^2)	184550	11140
(Ba, Cl^2)	194770	2070
(Na^2, O^2, SO^2)	257510	460
(Na^2, Cl^2)	195380	− 2360
(K^2, O^2, SO^2)	273560	− 6380
$(K^2, Cl^2,)$	211220	− 8880
(K^2, Cy^2)	130700	− 6020

The approximate general order of the metallic elementary substances, in which they evolve the most heat by chemical union with fluorine, oxygen, chlorine, bromine, acids, &c., is usually as follows:—Cs, Rb, K, Na, Li, Al, Ca, Ba, Sr, Mg, Zn, Tl, Cd, Sn, Pb, Fe, Co, Ni, Cu, Sb, Ag, As, Au, Pt. The following is their order when uniting with chlorine :—

Heat of Formation of Anhydrous Chlorides.

Metallic Element.	Compound formed.	Centigrade-Gramme Calories.
K	(K^2, Cl^2)	211220
Na	(Na^2, Cl^2)	195380
Ba	(Ba, Cl^2)	194740
Li	(Li^2, Cl^2)	187620
Sr	(Sr, Cl^2)	184550
Ca	(Ca, Cl^2)	169820
Mg	(Mg, Cl^2)	151010
Mn	(Mn, Cl^2)	111990
Al	$(Al^2, Cl^6) \times \cdot333$	107320
Zn	(Zn, Cl^2)	97210
Tl	(Tl^2, Cl^2)	97160
Cd	(Cd, Cl^2)	93240
Pb	(Pb, Cl^2)	82770
Fe	(Fe, Cl^2)	82050
Co	(Co, Cl^2)	76480
Ni	(Ni, Cl^2)	74530
Fe	$(Fe^2, Cl^6) \times \cdot333$	64027
Sn	$(Sn, Cl^4) \times \cdot5$	63625
Hg	(Hg, Cl^2)	63160
Bi	$(Bi, Cl^3) \times \cdot666$	60420
Ag	(Ag^2, Cl^2)	58760
Cu	(Cu, Cl^2)	51630
As	$(As, Cl^3) \times \cdot666$	47594
Sb	$(Sb, Cl^5) \times \cdot4$	41948
Te	$(Te, Cl^4) \times \cdot5$	38690
Au	$(Au, Cl^3) \times \cdot666$	15214

For more extensive extracts from Thomsen's tables see "Elements of Thermal Chemistry," by Muir and Wilson.

The next table gives the units of heat of chemical union of one gramme of metal with 8 grammes of oxygen or 35·5 grammes of chlorine. All the compounds formed were anhydrous except those indicated by a star.

Heat of Chemical Union.

Metals.	With Oxygen.	With Chlorine.
Potassium	76238*	100960
Sodium	73510*	94847
Zinc	42451	50296
Iron	37828	49651
Hydrogen	34462	23783
Lead	27675	44730
Copper	21885	29524
Silver	6113	34800

The determinations in the following table, being many of them older ones, and having been made by several different investigators, are less concordant than the previous ones. They are given in the order of the amounts of heat evolved by the combustion of one kilogramme of substance in oxygen expressed in centigrade-kilogramme calories at 0° to 1° C.

Heat of Formation of Oxides.

Substance.	Compound formed.	Centigrade-Kilogramme Calories.	Observer.
Hydrogen	Liquid H^2O	34180·	Thomsen
Carbon	CO^2	8080·	Favre and Silbermann
Silicon	SiO^2	7830·	Thomsen
Magnesium	MgO	6077·5	,,
Phosphorus	P^2O^5	5964·5	,,
Carbon	CO	2473·	Favre and Silbermann
Sulphur	SO^2	2221·3	Thomsen
Manganese	MnO^2	2113·0	,,
Iron	Fe^2O^3	2028·	Favre and Silbermann
Potassium	K^2O	1745·	Woods
Manganese	MnO	1724·	Thomsen
Iron	FeO	1352·6	Favre and Silbermann
Zinc	ZnO	1314·3	Thomsen
Tin	SnO^2	1147·	Andrews
,,	SnO	573·6	,,
Lead	PbO	243·	Thomsen
Mercury	HgO	153·5	,,
,,	Hg^2O	105·5	,,
Bismuth	Bi^2O^3	95·5	Woods
Silver	Ag^2O	27·3	Thomsen

SECTION C.

CHIEF CHEMICAL FACTS AND PRINCIPLES OF THE SUBJECT.

As in every electrolytic and voltaic action, substances either unite together chemically to form compounds, or compounds are chemically decomposed, it is manifest that every student of electro-metallurgy, and every person who superintends the electrolytic separation and refining of metals, requires some training in chemistry. He might with great advantage have previous experience in chemical and electrical manipulation, and possess a knowledge of the leading physical and chemical properties of the common elementary substances and their chief compounds ; of metals and metalloids, alkalies, bases and salts ; of chemical nomenclature, the use of chemical symbols, notation, formulæ, equations, and schemes of decomposition ; the meaning of the terms chemical affinity, elementary substance, atom, molecule, atomic, molecular and equivalent weight, valency, specific gravity, &c. ; and be able to perform chemical analyses and make chemical calculations. Knowledge of the meanings of the terms monobasic, bibasic, and tribasic is also necessary in order to be able to judge respecting the true electrolytic equivalents of compounds, and the relative amounts of substances decomposed by a current. As the reader is supposed to have already received sufficient preliminary training to enable him to understand the subject of this book, only a very brief and incomplete outline will be given of the chief points of chemical knowledge relating to electro-metallurgy.

EXPLANATION OF CHEMICAL TERMS.—By *chemical affinity* is meant the kind of energy which causes two dissimilar substances to unite together *in certain definite proportions by weight*, and produce a third homogeneous substance, widely different in its properties from the originals ; its action is nearly always attended by evolution of heat. The proportions in which they chemically unite are usually those of their atomic or molecular weights, or some simple multiple of these.

An *elementary body* is a substance which we have never yet been able to decompose or separate into two substances. An *atom* is the smallest particle of an elementary substance which can exist in a chemically combined state. A *molecule* is a group of two or more atoms chemically united together, and is the ·smallest particle of an elementary or compound substance which can exist in a chemically free state; a molecule of an elementary substance in the gaseous state is usually composed of two atoms. A *mass* is a collection of molecules; its amount is usually defined by its weight, but more accurately by the amount of its mechanical energy when moving at a given velocity. By *atomic weight* is meant the relative weight of an atom, that of hydrogen being the unit. Of the actual weight of atoms we know but little; it is, however, extremely small. The following is a table of atomic weights :—

Symbols and Atomic Weights of Elementary Substances.

Name.	Symbol.	Atomic Weight.	Name.	Symbol.	Atomic Weight.
Aluminium	Al	27·3	Mercury	Hg	199·8
Antimony	Sb	122·0	Molybdenum	Mo	95·6
Arsenic	As	74·9	Nickel	Ni	58·6
Barium	Ba	136·8	Niobium	Nb	94·0
Beryllium	Be	9·0	Nitrogen	N	14·01
Bismuth	Bi	210·0	Osmium	Os	198·6
Boron	Bo	11·0	Oxygen	O	15·96
Bromine	Br	79·75	Palladium	Pd	106·2
Cadmium	Cd	111·6	Phosphorus	P	30·96
Caesium	Cs	133·0	Platinum	Pt	196·7
Calcium	Ca	39·9	Potassium	K	39·04
Carbon	C	11·97	Rhodium	Ro	104·5
Chlorine	Cl	35·37	Rubidium	Rb	85·2
Cerium	Ce	141·2	Ruthenium	Ru	103·5
Chromium	Cr	52·4	Selenium	Se	78·0
Cobalt	Co	58·6	Silver	Ag	107·66
Copper	Cu	63·0	Silicon	Si	28·0
Didymium	Di	147·0	Sodium	Na	22·99
Erbium	Er	169·0	Strontium	Sr	87·2
Fluorine	F	19·1	Sulphur	S	31·98
Gallium	Ga	69·7	Tantalum	Ta	182·0
Gold	Au	196·2	Tellurium	Te	128·0
Hydrogen	H	1·0	Thallium	Tl	203·6
Indium	In	113·4	Thorium	Th	231·5
Iodine	I	126·53	Tin	Sn	117·8
Iridium	Ir	196·7	Titanium	Ti	48·0
Iron	Fe	55·9	Tungsten	W	184·0
Lanthanum	La	139·0	Uranium	Ur	240·0
Lead	Pb	206·4	Vanadium	Va	51·2
Lithium	L	7·01	Yttrium	Yt	93·0
Magnesium	Mg	23·94	Zinc	Zn	64·9
Manganese	Mn	54·8	Zirconium	Zr	90·0

The *molecular weight* of a substance is the sum of the relative weights of the atoms composing a molecule of it. The following are the relative molecular weights of some of the common substances, including those which are the most likely to be useful to the electrolytic chemist :—

Chemical Formulæ and Molecular Weights of Substances.

Substance.	Formulæ.	Molecular Weight.
Hydrogen	H^2	2·0
Oxygen	O^2	32·0
Fluorine	F^2	38·0
Chlorine	Cl^2	71·0
Bromine	Br^2	160·0
Iodine	I^2	254·0
Sulphur	S^2	64·0
Water	H^2O	17·96
Hydrofluoric acid	HF	20·0
Hydrochloric ,,	HCl	36·5
Carbonic ,,	CO^2	43·9
Nitric ,,	HNO^3	62·88
Hydrosulphuric ,,	H^2S	33·98
Sulphurous ,,	H^2SO^3	81·86
Sulphuric ,,	H^2SO^4	97·82
Phosphoric ,,	H^3PO^4	97·8
Arsenious ,,	As^2O^3	197·68
Arsenic ,,	As^2O^5	229·6
Terchloride of Antimony	$SbCl^3$	228·5
,, Bismuth	$BiCl^3$	316·5
Nitrate ,,	$Bi3NO^3+5H^2O$	485·44
Chloride of Platinum	$PtCl^4$	339·0
,, Gold	$AuCl^3$	303·0
Nitrate of Silver	$AgNO^3$	169·54
Fluoride ,,	AgF	126·66
Chloride ,,	AgCl	143·16
Iodide ,,	AgI	234·19
Mercuric chloride	$HgCl^2$	270·0
Cupric nitrate	$Cu2NO^3+6H^2O$	294·52
,, chloride	$CuCl^2$	133·74
,, sulphate	$CuSO^4+5H^2O$	249·5
Chloride of Nickel	$NiCl^2$	130·0
Sulphate ,,	$NiSO^4+7H^2O$	281·0
Nitrate of Cobalt	$Co2NO^3+6H^2O$	291·0
Chloride ,,	$CoCl^2$	130·0
Sulphate ,,	$CoSO^4+7H^2O$	281·0
Ferrous chloride	$FeCl^2$	127·0
Ferric ,,	Fe^2Cl^6	325·0
Ferrous sulphate	$FeSO^4+7H^2O$	278·0
Manganous chloride	$MnCl^2+4H^2O$	198·0
,, sulphate	$MnSO^4+5H^2O$	241·0

Chemical Formulæ, &c. (continued).

Substance.	Formulæ.	Molecular Weight.
Nitrate of Lead	$Pb2NO^3$	331·0
Sulphate of Thallium	Tl^2SO^4	504·0
Stannous chloride	$SnCl^2$	189·0
Chloride of Cadmium	$CdCl^2+2H^2O$	219·0
,, Zinc	$ZnCl^2$	136·0
Sulphate of Zinc	$ZnSO^4+7H^2O$	287·0
Chloride of Magnesium	$MgCl^2$	95·0
Sulphate ,,	$MgSO^4+7H^2O$	226·0
Cryolite	$6NaF, Al^2F^6$	421·
Chloride of Aluminium	Al^2Cl^6	268·
Sodio-chloride ,,	$2NaCl, Al^2Cl^6$	382·
Potash alum	$KAl2SO^4+12H^2O$	473·7
Chloride of Calcium	$CaCl^2$	110·
Caustic Lime	CaO	56·
,, Soda	$NaHO$	39·86
Chloride of Sodium	$NaCl$	58·4
Sulphate ,,	$Na^2SO^4+10H^2O$	321·44
Carbonate ,,	$Na^2CO^3+10H^2O$	285·48
Phosphate ,,	$Na^2 HPO^4+12 H^2O$	293·36
Caustic Potash	KHO	56·1
Nitrate of Potassium	KNO^3	101·
Chloride ,,	KCl	74·6
Chlorate ,,	$KClO^3$	122·5
Bromide ,,	KBr	119·1
Iodide ,,	KI	166·
Sulphate ,,	K^2SO^4	174·
Carbonate ,,	K^2CO^3	138·
Ammonia	H^3N	17·
Nitrate of Ammonium	H^4N, NO^3	80·
Chloride ,,	H^4N, Cl	53·5
Sulphate ,,	$(2H^4N)SO^4$	132·
Cyanogen	C^2N	26·
Hydrocyanic acid	HC^2N	27·
Cyanide of Potassium	KC^2N	65·
Oxalic acid	$H^2C^2O^4+2H^2O$	126·

An *equivalent weight* of a simple substance is either the atomic weight or some simple proportion of it, and is that weight which contains the same amount of chemical energy, or some simple proportion thereof, as the substance it is to combine with or decompose. An equivalent weight of a compound substance is either the molecular weight or some simple proportion of it. A substance may have several equivalent proportions, but can have only one atomic or molecular weight.

By *atomicity*, *valency*, or *atom-fixing power*, is meant the number of atoms of hydrogen or other monad element with which one atom of the elementary substance can combine, or which it can displace. A *monad* is an elementary substance, an atomic weight of which can combine with or displace one atom of a monad element ; a *dyad* is one, an atomic weight of which can combine with or displace 2 ; a *triad*, 3 ; a *tetrad*, 4 ; a *pentad*, 5 ; and a *hexad*, 6.

Degrees of valency always correspond to equivalent weights or combining proportions, whether these are the same as the atomic weights or not. As, however, there are instances in which the same elementary substance combines in several definite proportions with another, as in the case of nitrogen uniting with oxygen, chlorine with antimony, &c., difficulties have arisen in determining the true valencies of substances, and the following classification is therefore largely an arbitrary one:—

Monads.	Dyads.	Triads.	Tetrads.	Pentads.	Hexads.
Hydrogen	Oxygen	Boron	Carbon	Nitrogen	Molybde-
Fluorine	Sulphur	Rhodium	Silicon	Phosphorus	num
Chlorine	Selenium	Gold	Titanium	Arsenic	Vanadium
Bromine	Tellurium	Bismuth	Tin	Antimony	Tungsten
Iodine	Barium	Aluminium	Zirconium		Osmium
Caesium	Strontium	Indium	Thorium		Chromium
Rubidium	Calcium		Cerium		Manganese
Potassium	Lanthanum		Nickel		
Sodium	Didymium		Cobalt		
Lithium	Glucinum		Uranium		
Thallium	Magnesium		Lead		
Silver	Zinc		Ruthenium		
	Cadmium		Iridium		
	Iron		Palladium		
	Copper		Platinum		
	Mercury				

A *metal* is an elementary substance which conducts heat and electricity freely, and unites chemically with metalloids ; a *metalloid* is one of the opposite properties to these, and unites chemically with metals. An *alloy* is a homogeneous mixture or compound of metals. An *acid* is a compound substance, which chemically combines with and neutralises an alkali, tastes sour, and turns blue litmus-paper red ; it is usually soluble in water, and is an oxide of a metalloid or a peroxide of a metal. An *alkali* is a compound body, which chemically combines with and neutralises an acid, tastes soapy, and turns red litmus-paper blue; it is soluble in water, and is usually an oxide of a metal. A *base* is a compound substance, which by chemically uniting with an acid forms a salt; it is usually a metallic oxide, sometimes an alkali. A *salt* is a chemical compound, either of a metal and metalloid or of an acid and a base. An *anhydride* is a compound substance which by uniting with water forms an acid; it is usually an oxide of a metalloid. *Anhydrous* means destitute of water. *Hydrated* means containing water combined in definite proportions by weight.

Chemical Potentiality.—By this term is meant the amount of chemical energy stored up in a static or inactive state in substances under certain conditions, and ready to become dynamic or active under other circumstances, and converted into some other form of energy, such as heat, electric current, &c. It is of two kinds, viz., electro-negative, or that stored up in metalloids and acids; and electro-positive, or that contained in metals and bases. The former is greatest in amount in fluorine, oxygen, chlorine, &c., and the latter in the alkali-metals, potassium, sodium, &c.

Every substance, whether elementary or compound, has stored up in it a greater or less amount of potential chemical energy, and when substances unite together chemically, each loses a portion of its energy, and when they are again separated the amounts of energy lost are restored to them. The amount of potential energy in a substance may, in many cases, be liberated in several fractional portions in succession; for instance, if sodium be exposed to oxygen, it is slowly converted into oxide with liberation of a portion of its heat; if that oxide is exposed to contact with water it is changed into hydrate, with liberation of a second portion of heat; if the hydrate is brought into contact with carbonic anhydride gas it loses a third portion of energy, with formation of carbonate of sodium; if the carbonate is mixed with hydrochloric acid, a fourth portion of energy escapes, and chloride of sodium is formed; if that chloride is mixed with sulphuric acid, it is converted into sulphate with a fifth loss of heat, and if that sulphate is added to water it dissolves and loses a sixth portion of heat. These losses of heat are attended by losses of voltaic energy (*see Proc.* Birm. Phil. Soc., Vol. VI., Part 2).

Relation of Heat to Chemical Action.—In each case of chemical union, that compound is usually the most readily produced, the formation of which is attended by the liberation of the most heat; for instance, carbon in contact with oxygen usually burns to carbonic acid, evolving 48,465 centigrade-gramme calories, and not to carbonic oxide, evolving only 29,000. That substance also is usually the most readily corroded, which evolves the most heat by corrosion; thus, phosphorus rapidly oxidises in the air, whilst silver is unaffected, and the proportions of heat evolved by their oxidation are 73,980 and 5,900 thermal units respectively. With magnesium, zinc, and copper, in dilute sulphuric acid, magnesium is attacked and dissolved the most rapidly and copper the most slowly, and the respective amounts of heat evolved are 231,230, – 158,990, and 111,490 units.

In each case of chemical decomposition also those substances are usually the most easily separated which absorb the least amount of heat; thus, protosalt is more readily separated than metal from persalt of iron. The order in which substances are the most easily disunited is, as might be supposed, the reverse of that in which they most readily unite; thus whilst platinum, gold, and silver are very easily obtained from their compounds, potassium, sodium, magnesium, and the other alkali and earth-metals are much more difficult to liberate. In a solution, therefore, containing several dissolved metals, that one is usually the first to separate and appear which absorbs the least heat; thus, if a piece of zinc is immersed in a solution of the mixed chlorides of platinum, gold, mercury, copper, and nickel, platinum is the first and nickel the last precipitated.

One substance will usually only chemically displace and separate another, provided it evolves more energy in combining than the others absorb when separating. A metal will usually separate from their soluble compounds all other metals which require less energy to liberate them; for instance, zinc will precipitate from their chloride solutions all those which are below it in the tables given on pages 41 and 42. In like manner fluorine will expel all or nearly all other metalloids. Both in cases of chemical union and of decomposition those changes first occur to which there exist the smallest amounts of resistance.

E

SECTION D.

CHIEF FACTS OF CHEMICO-ELECTRIC OR VOLTAIC ACTION.

As voltaic action is the converse of electro-chemical or electrolytic action it is described somewhat more fully. In all cases of electrolysis voltaic influence has to be considered, because the electrolytic products give rise to polarisation due to voltaic action, which produces counter-electromotive force in the electrolysis vessel and tends to stop the original current. The degree of energy of this opposing voltaic influence depends upon that of the liberated electro-positive and electro-negative substances, and is indicated by the relative positions of those bodies in the following table ; the farther the two substances are apart in the series the more powerful usually is the opposing influence, and the stronger the voltaic current the two substances can generate :—

Chemico-Electric or Volta-Tension Series.

Caesium +	Magnesium	Nickel	Hydrogen	Rhodium	Selenium
Rubidium	Aluminium	Thallium	Mercury	Platinum	Phosphorus
Potassium	Chromium	Indium	Silver	Osmium	Sulphur
Sodium	Manganese	Lead	Antimony	Silicon	Iodine
Lithium	Zinc	Cadmium	Tellurium	Carbon	Bromine
Barium	Gallium	Tin	Palladium	Boron	Chlorine
Strontium	Iron	Bismuth	Gold	Nitrogen	Oxygen
Calcium	Cobalt	Copper	Iridium	Arsenic	Fluorine –

In this series each elementary substance is usually chemico-electro-positive to all those below it, and negative to those above it ; the nearer a substance is to the upper end, the more positive it is, and the nearer to the lower end the more negative. As none of the substances are absolutely positive or negative, and the electrical conditions are only relative ones, the series cannot be divided into two groups, the one positive and the other negative ; it is, however, customary to call the metals, especially the alkaline and easily corrodible ones, positive, and the metalloids negative. As also, the position of each member of the series varies considerably under different conditions, the order is largely an arbitrary one; but it is nevertheless very useful as a general guide to the usually relative chemical, thermic, and voltaic behaviour of the substances.

Electrical Theory of Chemistry.—The above order is substantially the same as that in which Berzelius arranged the chemical elements in his "Electrical Theory of Chemistry." According to his theory, the chemical union of any two substances is an electrical act, *i.e.*, that during contact, previous to union, the one substance is relatively positive and the other relatively negative, and the act of union is a consequence of the attraction existing between substances in these two states; also that during the act of uniting the two electric conditions neutralise each other and produce heat.

This theory may be reasonably extended from the union between metals and acids, or other conductors of electricity, to that between all non-conducting substances, because resistance to conduction is relative and not infinite. If, therefore, the electric attraction is sufficiently strong and the circuits sufficiently small to reduce the resistance to a minimum, chemical union and electric conduction between comparatively non-conducting substances must occur. The separation of gold, silver, and even copper, from aqueous solutions of their salts by simple immersion of a piece of ordinary white phosphorus in the liquids are good examples of electrolysis produced by a highly non-conducting substance; and the separation of hydrogen from pure water, by mere simple immersion in it of a magnesium platinum couple, is an instance of electrolysis in a comparatively non-conducting liquid. The chemical union of two non-conducting substances may therefore be reasonably regarded as only a somewhat more extreme example of the same kind of action occurring in cases where the resistance to conduction is reduced to a minimum by the shortness of the circuits.

Order of Voltaic Potentiality.—The order of the voltaic series is identical with that of chemical potentiality of the elements (p. 48), the positive or basic potentiality increasing towards the positive end, and the negative or acid potentiality towards the negative one. The order is also substantially the same as that of the amounts of heat evolved by the chemical union of the elements (*see* pp. 39-42). The substances nearest the ends of

E 2

the series exhibit the greatest, and those near the middle the
least electric, chemical, and thermal energy. The farther any
two of them are asunder in the order the greater usually is the
difference in their electric state, the stronger is the electromotive force of current they produce as a voltaic couple, the
greater is their chemical affinity for each other, the larger is
the amount of heat they evolve when chemically uniting, and
the greater is the quantity of heat or electric energy required
to decompose their compounds.

From a knowledge of this order, and of these general truths
involved in it, various phenomena may be predicted, and infer-
ences may be drawn respecting the probable effects in a great
number and variety of experiments ; thus—If A is positive to
B, and B is positive to C, A will be positive to C. If A evolves
more chemical heat than B, and B evolves more than C, A will
evolve more than C. If A displaces or separates B, and B displaces or separates C, A will displace or separate C. Or if A
has more potential energy and electromotive force than B, and
B has more than C, A will have more than C, and so on.

Volta-Tension Series of Electrolytes.—For a general order
of electrolytes, showing the relative degrees of electric energy
produced by them with a zinc platinum couple in a "voltaic
balance," *see Proc.* Birm. Phil. Soc., 1889-90, p. 43.

Relation of Chemical Heat to Volta-Motive Force.—Chemical heat and potentiality are intimately connected with volta-
motive force. In a voltaic cell each metal corrodes and evolves
heat, and tends to produce a current, the one in an opposite
direction to the other ; and it is the difference between these
two opposing forces which determines the direction of the current
and the degree of its electromotive force. Sir W. Thomson, in
1851, showed how to calculate the electromotive force of a
voltaic couple from thermo-chemical data, viz., by dividing the
difference of their amounts of chemical heat by about 23,000.
In a Daniell's cell zinc is corroded, and forms sulphate of zinc
which dissolves at the positive plate, and copper is separated
from its sulphate at the negative plate. The first of these
actions *evolves* $\dfrac{177420}{2} = 88710$ centigrade calories for each chemical equivalent in grammes of zinc dissolved, and the second

absorbs $\dfrac{127290}{2} = 63645$ such calories for each equivalent in grammes of copper set free, thus leaving a difference of 25,065 evolved at the zinc; and as the energy of about 22,900 such calories equals one volt, the electromotive force of the cell is

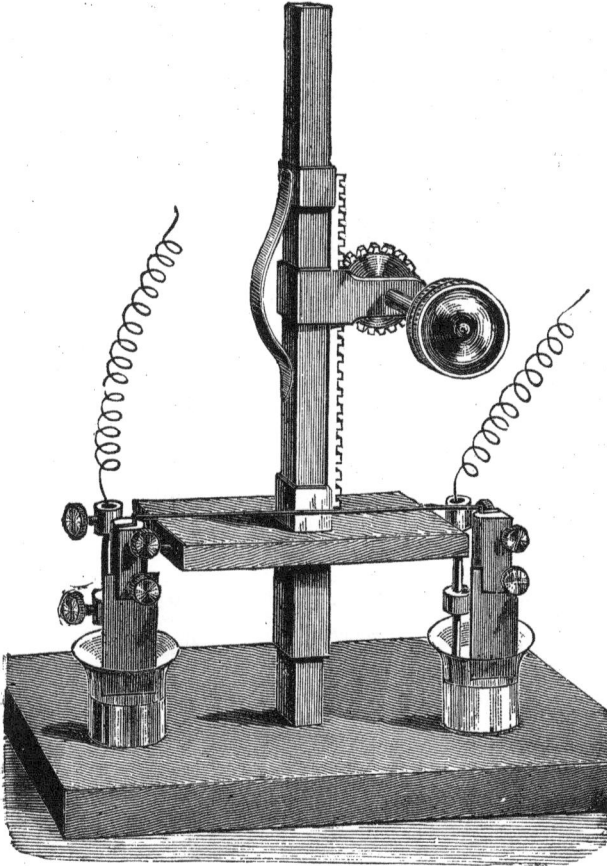

FIG. 13.—Voltaic Balance.

about 1·093 volt. As the true voltaic and electrolytic equivalent of a metal is that proportion of it which unites with one atomic weight of chlorine, or half an atomic weight of oxygen, the electromotive force must be calculated from that equivalent.

The following are the amounts of electromotive force generated by the chemical action of dilute hydrochloric acid on various metals, as thus calculated from the quantities of heat evolved by their chemical union with and solution in that liquid :—

Volta-Electromotive Forces.
As Calculated from Heat of Chemical Action and Solution.

Thermo-Chemical Formulæ.	Centigrade-Gramme-Calories.		Volts.
(K^2, Cl^2, Aq)	$\dfrac{202340}{2}$	$= 101070$	$= 4\cdot4135$
(Na^2, Cl^2, Aq)	$\dfrac{193020}{2}$	$= 96510$	$= 4\cdot2145$
(Mg, Cl^2, Aq)	$\dfrac{186930}{2}$	$= 93465$	$= 4\cdot0812$
(Al^2, Cl^6, Aq)	$\dfrac{475650}{6}$	$= 79275$	$= 3\cdot4616$
(Zn, Cl^2, Aq)	$\dfrac{112840}{2}$	$= 56420$	$= 2\cdot4638$
(Fe, Cl^2, Aq)	$\dfrac{99950}{2}$	$= 49975$	$= 2\cdot1821$
(Fe^2, Cl^6, Aq)	$\dfrac{255440}{6}$	$= 42573$	$= 1\cdot8590$
(Cd, Cl^2, Aq)	$\dfrac{96250}{2}$	$= 48125$	$= 2\cdot1011$
(Co, Cl^2, Aq)	$\dfrac{94820}{2}$	$= 47410$	$= 2\cdot0703$
(Ni, Cl^2, Aq)	$\dfrac{93700}{2}$	$= 46850$	$= 2\cdot0458$
(Sn, Cl^2, Aq)	$\dfrac{81140}{2}$	$= 40570$	$= 1\cdot7716$
(Tl^2, Cl^2, Aq)	$\dfrac{76960}{2}$	$= 38480$	$= 1\cdot6803$
(Pb, Cl^2, Aq)	$\dfrac{75970}{2}$	$= 37985$	$= 1\cdot6594$
(Cu, Cl^2, Aq)	$\dfrac{62710}{2}$	$= 31355$	$= 1\cdot3690$
(Hg, Cl^2, Aq)	$\dfrac{59860}{2}$	$= 29930$	$= 1\cdot3087$
(Te, Cl^4, Aq)	$\dfrac{97720}{4}$	$= 24430$	$= 1\cdot0668$
(Au, Cl^3, Aq)	$\dfrac{27270}{3}$	$= 9090$	$= \cdot3969$

According to these numbers a potassium-gold couple in dilute hydrochloric acid would yield a current having an electromotive force of about $4\cdot4135 - \cdot3969 = 4\cdot0166$ volts.

These and other similar data are useful in ascertaining the degree of polarisation which may be caused by the products of electrolysis whilst electrolysing a liquid. For instance, in electrolysing dilute sulphuric acid with platinum electrodes oxygen is set free at the anode and hydrogen at the cathode, and by their chemical reunion evolve $\dfrac{68360}{2}$ centigrade-gramme calories per each one gramme of hydrogen recombining, and tend to cause a counter electromotive force of 1·492 volt.

It must, however, be remarked that this mode of calculating the electromotive force of a voltaic couple fails in various cases. This has been observed by Wright and Thompson (*Phil. Mag.*, Vol. XVII., p. 378, 1884 ; *Proc.* Roy. Soc., Vol. XLIV., p. 197, 1888) ; by E. F. Herroun (*Phil. Mag.*, Vol. XXI., p. 20, 1886), and by others.

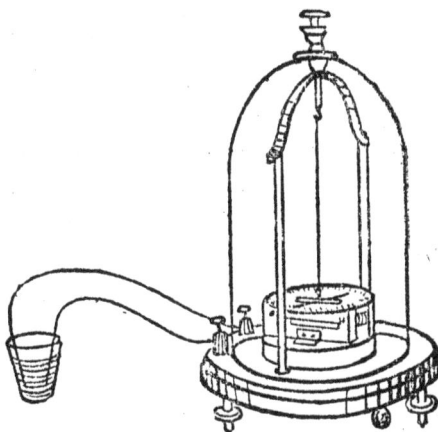

FIG. 14.—Voltaic Couple and Galvanometer.

The relative positions of any two metals in the volta-tension series, with any given electrolyte, may be experimentally ascertained by connecting them with the two ends of the coil of a galvanometer, and simultaneously immersing their free ends in the liquid ; an electric current is then produced, proceeding from the more positive metal through the liquid to the negative one, and its direction is shown by that of the movement of the galvanometer (Fig. 14) needles. Or if an electrolyte contains in solution the salts of two easily reducible metals, and a piece of each of those metals be separately immersed in it, the least positive, and consequently the most easily reducible one, will be deposited as a coating upon the other. The rela tive degrees of electromotive force of any two voltaic couples may be measured by the methods given on p. 15.

Volta-Electric Relations of Metals in Electrolytes.—The order of volta-electromotive force differs to some extent in every different electrolyte, and numerous experiments have been made by various investigators in this subject; some of the results obtained are given in the following series :—

1. *In Dilute Acids* (H. Davy) + Potassium, Barium, Zinc amalgam, Zinc Ammonium-amalgam, Cadmium, Tin, Iron, Bismuth, Antimony, Lead, Copper, Silver, Palladium, Tellurium, Gold, Platinum, Iridium, Rhodium. –

2. *Equal volumes of Sulphuric Acid and Water* (Pfaff) + Zinc, Cadmium, Tin, Lead, Tungsten, Iron, Bismuth, Antimony, Copper, Silver, Gold, Tellurium, Platinum, Palladium. –

3. *Very Dilute Sulphuric Acid* (Marianini) + Zinc, Charcoal (heated and quenched), Clean Lead, Tin, Manganese, Tarnished Lead, Iron, Magnetic Iron Ore, Brass, Copper, Rusty Brass, Bismuth, Nickel, Charcoal, Antimony, Tinstone, Native Sulphide of Molybdenum, Arsenic, Tarnished Antimony, Silver, Mercury, Tarnished Arsenic, Tarnished Arsenical Silver, Red Silver, Galena, Fresh Charcoal, Copper-nickel, Copper-glance, Arsenical Cobalt, Black Tellurium, Copper Pyrites, Platinum, Gold, Auriferous Native Tellurium, Cubical Iron Pyrites, Graphite, Arsenical Pyrites, Magnetic Pyrites, Amorphous Iron Pyrites, Peroxide of Manganese, Old Charcoal. –

4. *Hydrochloric Acid* (Faraday) + Zinc, Cadmium, Tin, Lead, Iron, Copper, Bismuth, Nickel, Silver, Antimony. –

5. *One volume of Nitric Acid and seven volumes of Water* (Faraday) + Zinc, Cadmium, Lead, Tin, Iron, Nickel, Bismuth, Antimony, Copper, Silver. –

6. *Nitric Acid*, sp. gr. 1·48 (Faraday) + Cadmium, Zinc, Lead, Tin, Iron, Bismuth, Copper, Antimony, Silver, Nickel. –

7. *Concentrated Nitric Acid* (De la Rive) + Tin, Zinc, Iron, Copper, Lead, Mercury, Silver, Peroxide of Iron. –

8. *Pure Dilute Hydrofluoric Acid of 10 per cent.* (Gore) + Aluminium, Zinc, Magnesium, Thallium, Cadmium, Tin, Lead, Silicon, Iron, Nickel, Cobalt, Antimony, Bismuth, Copper, Silver, Gold, Gas-carbon, Platinum. –

9. *Pure Dilute Hydrofluoric Acid of 28 per cent.* (Gore) + Zinc, Magnesium, Aluminium, Thallium, Indium, Cadmium, Tin, Lead, Silicon, Iron, Nickel, Cobalt, Antimony, Bismuth, Mercury, Silver, Copper, Arsenic, Osmium, Ruthenium, Gas-carbon, Platinum, Rhodium, Palladium, Tellurium, Osmi-iridium, Gold, Iridium. –

10. *Solution of Potash or Soda, strong or weak* (Faraday) + Zinc, Tin, Cadmium, Antimony, Lead, Bismuth, Iron, Copper, Nickel, Silver. –

11. *Aqueous Ammonia,* sp. gr. ·95 (Pfaff) + Zinc, Tin, Lead, Silver, Copper. –

12. *One part of Cyanide of Potassium in eight parts of Water* (Poggendorff) + Amalgamated Zinc, Zinc, Copper, Cadmium, Tin, Silver, Nickel, Antimony, Lead, Mercury, Palladium, Bismuth, Iron, Platinum, Cast-iron, Coke. –

13. *Dilute solution of yellow Sulphide of Potassium* (Faraday) + Zinc, Copper, Cadmium, Tin, Silver, Lead, Antimony, Nickel, Bismuth, Iron. –

14. *Colourless solution of Sulphide of Potassium* (Faraday) + Cadmium, Zinc, Copper, Tin, Antimony, Silver, Lead, Bismuth, Nickel, Iron. –

15. *Dilute Solution of Hydrosulphate of Potassium* (H. Davy) + Zinc, Tin, Copper, Iron, Bismuth, Silver, Platinum, Palladium, Gold, Charcoal. –

16. *Solution of Salammoniac* (Poggendorff) + Zinc, Cadmium, Manganese, Lead, Tin, Iron, Steel, Uranium, Brass, Magnetic-iron, German Silver, Cobalt, Bismuth, Antimony, Arsenic, Chromium, Silver, Mercury, Copper Pyrites, Tellurium, Gold, Galena, Coke, Platinum, Plumbago, Peroxide of Manganese. –

17. *Solution of Chloride of Sodium* (Fechner) + Zinc, Lead, Tin, Iron, Antimony, Bismuth, Copper, Silver, Gold, Platinum. –

18. *Fused Boracic Acid* (Gore) + Iron, Silicon, Carbon, Platinum, Gold, Copper, Silver. –

19. *Fused Phosphoric Acid* (Gore) + Zinc, Iron, Copper, Silver, Platinum. –

20. *Fused Potassic Hydrate* (Gore) + Silicon, Aluminium, Zinc, Iron, Lead, Carbon, Platinum, Silver. –

21. *Fused Potassic Carbonate* (Gore) + Silicon, Iron, Zinc, Carbon, Copper, Silver, Platinum. –

22. *Fused Potassic Chloride* (Gore) + Aluminium, Zinc, Iron, Copper, Silver, Platinum. –

23. *Fused Potassic Fluoride* (Gore) + Palladium, Gold, Platinum, Iridium. –

24. *Fused Ammonium Nitrate* (Gore) + Magnesium, Zinc, Lead, Copper, Silver, Tin, Aluminium, Iron, Silicon, Carbon, Platinum. –

25. *Fused Potassic Cyanide* (Gore) + Magnesium, Zinc, Aluminium and Platinum, Nickel, Iron, Silver, Iridium, Gold, Rhodium, Carbon, Palladium, Cobalt, Antimony. –

For additional tables of the electrical relations of metals in fused salts, see *Phil. Mag.*, June, 1864; *Chemical News*, Vol. IX., p. 266; also *Proc. Roy. Soc.*, Vol. XXX., 1879, p. 48.

The above series are only those obtained by immersing two pieces of different metal in one liquid, but others may be obtained by immersing two pieces of the same metal in two different liquids which touch each other through a thin porous partition, or by immersing two different metals in two different liquids.

Voltaic Batteries.—Voltaic batteries are combinations such as the above, and are arrived at by means of similar experiments, those being selected which possess the best combination of qualities, such as sufficient electromotive force, least internal resistance, greatest strength of current, cheapness, manageability, freedom from noxious fumes, &c. There are single fluid batteries and double fluid ones ; one of high electromotive force can only be obtained by selecting two metals considerably asunder in the voltaic series. All other circumstances being alike, the most rapidly corroded metal with the least corroded one usually yields the strongest current. The strongest batteries are usually those composed of two metals and two liquids. Increasing the size of a voltaic cell increases the strength or quantity of the current per unit of time, by diminishing the internal resistance, but does not affect the electromotive force. The latter depends almost entirely upon the particular combination of metals and liquids employed, and essentially upon their heats of chemical combination (*see* pp. 52-54). The following are the electromotive forces of several useful batteries :—

Electromotive Forces of Voltaic Batteries.

Kind of battery.	Electromotive force. Volts.		
Smee	·64	very variable.	
Pabst	·78		
Wollaston	·8	to	1·04
Poggendorff	1·77	,,	2·27
Daniell	1·03	,,	1·12
Leclanché	1·4	,,	1·6
L. Clark's Standard	1·438		
Marie-Davey	1·5		
Bunsen	1·73	to	1·94
Grove	1·76	,,	1·93
Secondary cells.			
Planté, Faure, &c.	1·96	,,	2·22

Wollaston's battery is one of the oldest kinds; it consists of zinc and copper plates immersed in dilute sulphuric acid; it is not much used (Fig. 15).

Smee's is composed of two plates of amalgamated zinc, with a thin sheet of platinised silver between them. The use of the coating of platinum upon the silver is to prevent the adhesion of hydrogen gas, which largely diminishes the current (Fig. 16).

Poggendorff's consists of plates of zinc and carbon in a mixture of sulphuric acid 1 part, bichromate of potash 2 parts, and water 12 parts; the zinc being raised out of the liquid when the cell is not in action (Fig. 17).

Pabst's is composed of plates of iron and carbon in a solution of perchloride of iron. If the current taken from it is weak, the solution regenerates itself by absorbing oxygen from the air, otherwise it must be occasionally boiled after addition of some nitric acid.

Daniell's consists of amalgamated zinc in dilute sulphuric acid and copper in a nearly saturated solution of cupric sulphate, the two liquids being separated by a porous partition of baked earthenware. The upper part of the cupric solution is kept

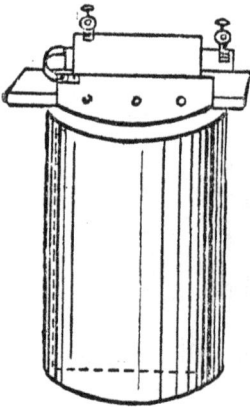

FIG. 15.—Wollaston's Cell.   FIG. 16.—Smee's Cell.

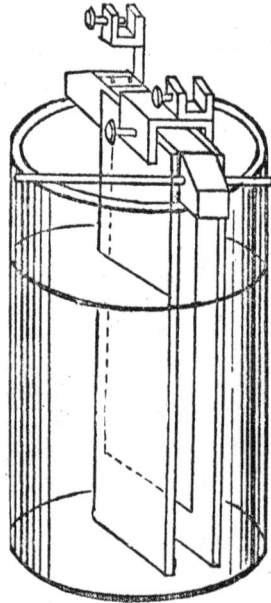

saturated by contact with crystals of blue vitriol. This battery is very little liable to polarisation, and the current it yields is very constant in strength (Fig. 18).

Leclanché's is excited by a solution of salammoniac, the liquid being divided into two parts by a porous partition. One division contains a bar of zinc, and the other a plate of carbon coated with a paste of manganese binoxide to prevent polarisation (Fig. 19).

Latimer Clark's Standard Cell is composed of a piece of zinc in contact with a wet paste of mercurous sulphate lying upon

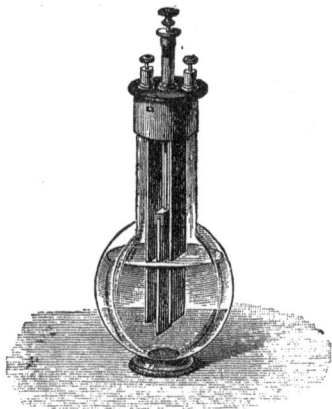

FIG. 17.—Poggendorff's Cell.　　FIG. 18.—Daniell's Cell.

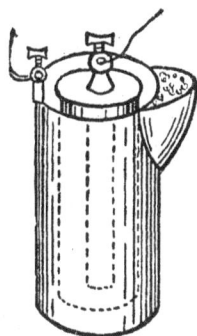

mercury in contact with an insulated platinum wire, which is used as the positive pole.

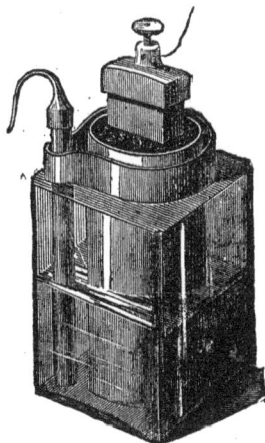

FIG. 19.—Leclanché's Cell.

Grove's consists of amalgamated zinc in dilute sulphuric acid, and a sheet of platinum in strong nitric acid, with an earthen porous partition separating the liquids (Fig. 20). It is a powerful battery, has small internal resistance, and the nitric acid prevents polarisation by oxidising the separated hydrogen. Its chief disadvantage is the acid fume it emits.

Bunsen's differs only from Grove's in a bar of gas carbon being substituted for the platinum.

For electrolytic experiments a Daniell's or a Smee's battery is as convenient as any, with plates as large or larger than the electrodes in the electrolysis cell, and having cells containing a copious stock of the exciting liquids.

As voltaic batteries are not much used in the commercial separation or refining of metals, and the number of kinds of

FIG. 20.—Grove's Battery.

batteries is large, the reader is referred for further particulars to books more specially devoted to them.

Influence of Temperature and Strength of Liquid on the Voltaic Order of Metals.—The order of volta-motive force in liquids, varies somewhat in the same kind of liquid with every different strength of liquid and every different temperature. Series of experiments upon the influence of these circumstances, with solutions of chloride, bromide, iodide, and cyanide of potassium, have been made by the author. The results of them are published in *Proc.* Roy. Soc., Vol. XXX., pp. 38-49, from which the following fragmentary tables are extracted. The amount of distilled water used with each quantity of salt in the first three tables was 50 cubic centimetres, and in Table IV. it was 800 grains :—

TABLE No. 1.—Potassic Chloride.

At 55°F. Grains.		At 100°F. Grains.	
5.	240.	5.	240.
Mg	Mg	Mg	Mg
Zn	Zn	Zn	Zn
Tl	Cd	Tl	Cd
Al	Al	Al	Al
Cd	Tl	Cd	Tl
In	In	In	In
Pb	Pb	Pb	Pb
Fe	Sn	Fe	Sn
Sn	Fe	Sn	Fe
Co	Co	Co	Co
Si	Si	Si	Sb
Sb	Sb	Sb	Cu
Bi	Cu	Bi	Bi
Cu	Bi	Cu	Si
Ni	Ni	Ni	Ni
Te	Ag	Te	Ag
Ir	Te	Ir	Te
Hg	Hg	Hg	Hg
Rh	Ir	Rh	Ir
Pd	Au	Pd	Au
Au	Pd	{ Au	Pd
Ag	Pt	Ag	Rh
Pt	Rh	Pt	Pt
C	C	C	C

TABLE No. 2.—Potassic Bromide.

At 55°F. Grains.		At 100°F. Grains.	
5.	360.	5.	360.
Mg	Mg	Mg	Mg
Zn	Zn	Zn	Zn
Tl	Cd	Tl	Cd
Cd	Ga	Cd	Tl
In	Tl	In	Al
Al	Al	Al	In
Fe	In	Pb	Pb
Pb	Pb	Fe	Sn
Sn	Sn	Sn	Fe
Co	Fe	Co	Cu
Sb	Cu	Sb	Co
Si	Co	Si	Sb
Bi	Sb	Bi	Si
Ni	Si	Ni	Bi
Cu	Bi	Cu	Ag
Te	Ag	Te	Ni
Hg	Ni	Hg	Hg
Ag	Hg	Ag	Pd
Pd	Pd	Pd	Au
Ir	Au	Ir	Te
Au	Te	Au	Pt
Rh	Pt	Rh	Ir
Pt	Ir	Pt	Rh
C	Rh	C	C
	C		

TABLE No. 3.—Potassic Iodide.

At 60°F. Grains.		At 100°F. Grains.	
5.	360.	5.	360.
Mg	Mg	Mg	Mg
Zn	Zn	Zn	Zn
Cd	Cd	Cd	Cd
Ga	Ga	Tl	Tl
Tl	Tl	Al	Al
Al	Al	Pb	Pb
Pb	Pb	Sn	Sn
Sn	Sn	In	In
In	In	Fe	Fe
Fe	Fe	Co	Cu
Co	Cu	Sb	Ag
Sb	Ag	Cu	Sb
Cu	Sb	Si	Co
Si	Co	Ag	Hg
Bi	Hg	Ni	Bi
Ag	Bi	Bi	Si
Ni	Si	Hg	Ni
Hg	Ni	Pd	Pd
Pd	Pd	Au	Au
Au	Au	Te	Te
Te	Te	Pt	Pt
Pt	Pt	Ir	Ir
Ir	Ir	Rh	Rh
Rh	Rh	C	C
C	C		

TABLE No. 4.—Potassic Cyanide.

At 50°F. Grains.		At 100°F. Grains.	
5.	240.	5.	240.
Al	Mg	Mg	Mg
Mg	Zn	Zn	Al
Zn	Cu	Cd	Zn
Cu	Al	Al	Cu
Cd	Cd	Co	Cd
Sn	Au	Cu	Pd
Co	Ag	Ni	Sn
Ni	Pd	Sn	Au
Ag	Ni	In	Ag
In	Sn	Au	Ni
Au	Hg	Tl	Hg
Tl	Pb	Ag	Pb
Hg	Co	Pb	Tl
Pb	Sb	Pd	Co
Si	Tl	Hg	Sb
Fe	In	Si	In
Pt	Te	Sb	Te
Pd	Bi	Bi	Bi
Sb	Si	Fe	Fe
Bi	Fe	Te	Si
Ir	Pt	Ir	Pt
Te	Ir	Pt	Ir
Rh	Rh	Rh	Rh
C	C	C	C

Dr. Gross has discovered that when iron is magnetised it becomes more electro-positive in liquids, and Nichols and Richards have verified this, and shown that it also evolves more chemical heat when dissolving in acids.

Relative Amounts of Voltaic Current Produced by Different Metals.—Equal weights of different metals in dissolving yield unequal amounts of current; if, however, the metals are taken in the proportions of their atomic weights, a very different relation is observed. If we call the amount of current yielded by one atomic weight of a monad metal (*see* pp. 46 and 47) 1, that yielded by an atomic weight of a dyad is 2, by a triad 3, a tetrad 4, and so on. The amount of current therefore generated by a given weight of a particular metal depends upon its atomic weight and valency.

Table of Voltaic Equivalents of Elementary Substances.

Substance.	Atomic Weight.	Equivalent Weight.	Coulombs per Gramme.
Aluminium	27·3	9·1	10615·7
Antimony	122·0	40·66	2415·4
Bromine	79·75	79·75	1207·4
Cadmium	111·6	55·8	1731·6
Calcium	39·9	19·95	4835·6
Carbon	11·97	3·	32206·1
Chlorine	35·5	35·5	2721·1
Cobalt	58·6	29·3	3274·4
Copper	63·0	31·5	3066·5
Fluorine	19·0	19·0	5086·4
Gold	196·2	65·4	1474·9
Hydrogen	1·0	1·0	96613·3
Iodine	126·53	126·53	760·6
Iron	55·9	27·95	3450·6
,, in per salts	55·9	18·64	5184·0
Lead	206·4	103·2	935·9
Lithium	7·01	7·01	13802·6
Magnesium	23·94	11·97	8645·2
Manganese	54·8	18·27	5288·2
Mercury	199·8	99·9	966·2
Nickel	58·6	29·3	3274·4
Nitrogen	14·0	4·3	22472·0
Oxygen	15·96	7·98	12121·2
Silicon	28·0	14·	6901·3
Silver	107·66	107·66	894·4
Sodium	22·99	22·99	4199·9
Sulphur	31·98	16·	6038·7
Tin	117·8	58·9	1640·4
Zinc	64·9	32·45	2972·6

And if we further take the metals in the proportions of their chemical equivalents, taking care in the case of those having several valencies or equivalents to select the one which unites with 1 atomic weight of chlorine or $\frac{1}{2}$ an atomic weight of oxygen, we find that a single chemical equivalent of each metal yields the same amount of current. Whilst the degree of volta-electromotive force of the current generated by a metal depends upon its quantity of chemical heat (see p. 54), the amount of current depends upon the chemical equivalent.

The table on page 64 shows the equivalent weights of the more common elementary substances required to produce the same amount of current, and the number of coulombs of current produced by one gramme of each substance.

Influence of Ordinary Chemical Corrosion.—The amount of current obtained, however, by actual experiment with different metals is very variable, and the full proportion is only secured in a few cases and under special conditions. This arises from the circumstance that a greater or less proportion of the metal is dissolved by ordinary chemical corrosion—especially in a hot liquid (see p. 69), the conduction resistance of which is so much diminished by the heat—in producing minute local currents, which circulate in myriads all over its surface, and do not enter the general stream or external circuit at all. By actual trial in nearly one hundred cases of various kinds I found that the proportion of external current obtained with electrolytes at ordinary temperature varied from 2 to nearly 100 per cent. ("On Some Relations of Chemical Corrosion to Voltaic Currents," *Proc.* Roy. Soc., Vol. XXXVI., 1884, pp. 331-340), and with hot liquids would no doubt be much less.

The following tables give a few of the cases with copper, and with zinc as the positive plate. The liquid employed was in each case divided into two equal portions of 3 or 4 ounces by measure in similar glass vessels, with a comparison plate wholly immersed in one portion and the positive plate of perfectly similar metal of equal size in the other portion, surrounded by a negative cylindrical sheet of platinum, and the current from this voltaic couple was caused to deposit silver by means of two electrodes of pure sheet silver in a separate solution of argento-cyanide of potassium in a third glass vessel.

F

The rates of loss of the positive plate and of the comparison one are expressed in grains per square inch per hour :—

COPPER AS POSITIVE METAL IN	Sq. in.	Hrs.	E.M.F. in volts.	Loss by positive plate.	Loss by comparison plate.	Grains weight of silver d'p'sit'd	Per cent of equivalnt of current.
Grains. Of water.							
10 KCy (89·14 p.c.) +1 oz..	6	1	·9706	·240	·172	·123	15·05
100 ,, ,, + ,,	6	1	1·1758	·530	·190	·577	32·00
Minims.							
25 strong HNO³ + ,,	6	21	·4946	2·550	·763	5·420	62·45
50 ,, ,, + ,,	6	3	·6429	·746	·809	1·858	73·20
48 ,, HCl + ,,	4	19	·466	·510	·181	·463	26·65
160 ,, + ,,	4	18	·3982	·384	·125	·320	24·50
160 ,, HClO³ + ,,	6	4	·6872	1·416	1·109	·449	9·30
1·25,, H²SO⁴ + ,,	6	17	·3111	·170	·109	·079	13·65
48 ,, ,, + ,,	6	19	·2947	·150	·133	·156	21·30
Concentrated	3	7	·4729	·075	·017	·047	20·61
ZINC AS POSITIVE METAL IN							
Grains. Of water.							
100 KHO +1 oz..	4	4	1·243	1·620	·563	·904	16·88
10 KCy + ,,	3	1·5	1·085	·188	·051	·484	77·45
100 ,, + ,,	4	5	1·1435	1·410	·391	2·996	64·00
Minims.							
1·25 strong HNO³ + ,,	3	·5	1·2726	·461	·343	·137	6·45
5· ,, HCl + ,,	3	1·	1·185	1·142	·482	·790	20·81
30· ,, ,, + ,,	4	·5	1·225	1·334	2·379	·354	8·10
1·25 ,, H²SO⁴ + ,,	3	·5	1·117	·389	·113	·283	21·9
10 ,, HF, SiF⁴ + ,,	3	2·	1·156	·399	·250	·639	48·7

Whilst silver and platinum as a voltaic couple in concentrated sulphuric acid gave 78·2 per cent. of equivalent of external current, and the positive plate lost only ·101 ; in a weak solution of ferric sulphate they gave only 1·3 per cent. equivalent of such current, and the positive plate lost 3·23.

The simple and chief explanation of the great variation in the proportion of corrosion by "local action" to that attended by external current appears to be difference of electric conduction resistance. When there is no external current, and the outer resistance is infinite, as in the case of deposition of metal or of hydrogen by simple immersion, all the electricity flows through the local circuits, and deposits its complete equivalent of metal, of hydrogen, or both, on the negative parts of the positive plate; but when the external resistance is least, and the resistance to circulation of local currents greatest, nearly the whole of the electricity passes through the outer circuit, and nearly all the deposition of hydrogen or metal occurs on the negative plate ; in either case the full equivalent of electricity circulates. With zinc and platinum in dilute sulphuric acid, the "local action" and evolution of hydrogen gas from the zinc, were diminished on closing the external circuit, thus showing the influence of the general current in protecting the zinc from ordinary chemical corrosion. By diminishing also the external resistance in proportion to that in the battery, by the employment of two voltaic cells in the series, instead of one, the proportion of silver (deposited in a separate electrolysis cell by the current) "was increased from 90·02 to 98·83 per cent. of the theoretical quantity."

Influence of External Resistance.—A given voltaic cell can only yield a certain maximum strength of current, and any conductor, not acting as a shunt introduced into the circuit, diminishes that amount by increasing the total resistance. The greater the electromotive force of the cell, the less is the strength of current diminished by increase of total resistance. If the external resistance is very small, an increase of electromotive force of the battery adds very little to the strength of the current; but if it is very large the opposite effect occurs. The difference of effect usually produced by means of a current from a single cell and that from many does not arise from any known difference in the nature of the current in the two cases, but usually from difference of proportion of internal to external resistance.

Influence of Kind of Substance on Chemical Corrosion.—Rate of corrosion depends chiefly upon the particular combination of metal and liquid. Thus, the relative rates with magnesium in solution of potassic cyanide (or in dilute hydrofluoric acid), and in dilute sulphuric acid, are just the opposite of those with aluminium in the same liquids. Whilst magnesium dissolves very slowly in the former liquid and very rapidly in the latter, aluminium does just the reverse; therefore, either a difference of metal or of liquid in this case reverses the effect (*see* Table, p. 69).

Influence of Temperature on Chemical Corrosion.—Rate of corrosion increases rapidly by rise of temperature. In two series of experiments with twelve different metals in various electrolytes at 60° and 160° F. respectively, I found "that in nearly every case, rise of temperature increased the rate of corrosion. The only exception in fifty-seven cases was zinc in dilute nitric acid, and that was a very feeble one. The increase of rate of corrosion was very variable : whilst that of tin in dilute nitric acid was increased 1·5 times, that of copper in chloric acid was increased 321·6 times ; both these metals and acids were very pure, and not a trace of oxide appeared upon the tin. The total average increase for the entire series was 29·98."

The tables on the next page exhibit the results obtained, stated in grains per square inch per hour. The proportion of substance dissolved in each ounce of distilled water to form the solutions were—of potassic chloride, bromide, iodide, sulphate, sodic chloride, potash alum, ammonia alum, potassic cyanide (containing 89·14 per cent. of actual cyanide), 10 grains each ; chloric acid, 6 minims ; nitric, hydrochloric, and sulphuric acids, 1¼ minims each. With silver, palladium, gold, and platinum it was necessary to use much larger sheets and to continue the experiment during a longer period. The solutions used with these metals, of the two alums were four times, and those of potassic cyanide, sulphuric, hydrochloric, nitric, and chloric acids, were of ten times the above strength. The action was in no case allowed to continue longer than was sufficient to exhaust about ten to twenty per cent. of the corroding substance. The silver was immersed in the cold dilute nitric and chloric acids during one month, and the palladium, gold, and platinum during three months. In the hot liquids the periods of immersion varied from five minutes to eight hours. The horizontal lines in the first table separate the weak solutions from the strong ones (*Proc.* Roy. Soc., Vol. XXXVII., pp. 283-287).

As in a voltaic cell, especially with a hot liquid, a large proportion of the positive metal may dissolve without producing external or useful current, so also conversely in an electrolysis vessel, especially with a warm electrolyte, much current may pass without permanently depositing metal (*see* pp. 121-124).

Influence of Unequal Temperature of the Metals.—It has long been known that if either one of the metals of a voltaic couple is singly heated the current is altered in strength, and

Corrosion Series of Metals at 60° F.

	1 K Cy	2 H² SO⁴	3 H Cl	4 H NO³	5 K Cl	6 Am Alum	7 H Cl O³	8 K Br	9 K Alum	10 K I	11 Na Cl	12 K² SO⁴
1	Al ·0095	Mg 1·0000	Mg ·1581	Zn ·6130	Cd ·0036	Mg ·13770	Mg ·3164	Cd ·00058	Mg ·00380	Zn ·00025	Fe ·00026	Pb ·00060
2	Zn ·0054	Fe ·0189	Zn ·0490	Cd ·3780	Zn ·0030	Zn ·00101	Fe ·1068	Zn ·00050	Zn ·00047	Fe ·00020	Pb ·00012	Fe ·00048
3	Cu ·0053	Cd ·0026	Fe ·0118	Sn ·3750	Fe ·0040	Cd ·00066	Sn ·0480	Cu ·00029	Fe ·00046	Cd ·00015	Cu ·00006	Zn ·00040
4	Ag ·0016	Pb ·0018	Cd ·0026	Mg ·1570	Sn ·0025	Pb ·00060	Ni ·0079	Fe ·00026	Cd ·00041	Ag ·00005	Sn ·00005	Sn ·00036
5	Cd ·0014	Sn ·0009	Pt ·0023	Fe ·1406	Cu ·0012	Fe ·00059	Pb ·0074	Pb ·00022	Ni ·00022	Sn trace	Ni ·00001	Cd ·00028
6	Pd ·0012	Cu ·0006	Cu ·0018	Pb ·0250	Al ·0006	Cu ·00038	Cd ·0040	Al ·00010	Cu ·00018	Ni less	Ag trace	Ag none
7	Au ·0009	Ni ·0005	Sn ·0017	Ni ·0030	Ni trace	Ni ·00025	Cu ·0008	Ni none	Sn ·00005	Pd none	Au none	Au ”
8	Sn ·0006	Al ·0004	Ni ·0004	Cu ·0008	Ag less	Sn ·00013	Ag ·000026	Au ”	Ag ·00000032	Au ”	Pd ”	Pd ”
9	Ni ·0005	Pd none	Pd none	Ag ·00022	Pd none	Ag ·00000032	Au none	Pd ”	Pd none	Pt ”	Pt ”	Pt ”
10	Fe ·0042	Au ”	Au ”	Pd none	Au ”	Au none	Pd ”	Pt ”	Au ”			
11	Pt ·0005	Pt ”	Pt ”	Au ”	Pt ”	Pd ”	Pt ”		Pt ”			
12	Pt ·00000036			Pt ”		Pt ”						

Corrosion Series of Metals at 160° F.

	1 K Cy	Increase from 1 to	2 H Cl	Increase from 1 to	3 H Cl O³	Increase from 1 to	4 Am Alum	Increase from 1 to	5 H NO³	Increase from 1 to	6 H² SO⁴	Increase from 1 to	7 K Alum	Increase from 1 to
1	Al ·4940	52·	Mg 1·2106	7·65	Mg 1·4507	4·58	Mg ·3520	2·55	Mg ·8853	5·64	Mg 4·0426	4·04	Mg ·1226	32·21
2	Cu ·1258	23·7	Zn ·2090	4·26	Fe ·6140	5·84	Fe ·0732	124·00	Cd ·8176	2·16	Zn 2·510		Zn ·0821	178·00
3	Zn ·0495	9·1	Fe ·1370	11·6	Cu ·2573	321·6	Zn ·0382	37·80	Zn ·6040	·98	Fe 2·493	13·20	Fe ·0386	82·10
4	Au ·0282	31·3	Al ·0273	10·0	Zn ·2160		Pb ·0138	23·00	Fe ·6000	4·26	Al ·284	71·00	Pb ·0138	
5	Ag ·0070	4·4	Pb ·0231	8·66	Pb ·1940	26·20	Cd ·0073	11·00	Sn ·5666	1·51	Pb ·0196	10·90	Al ·0090	
6	Pd ·0051	4·2	Cu ·0156	3·7	Sn ·1660	3·52	Al ·0071		Pb ·2155	8·62	Sn ·0115	12·80	Sn ·0080	160·0
7	Cd ·0046	3·3	Cd ·0096	4·6	Cd ·1070	26·70	Sn ·0060	46·10	Al ·0260		Cd ·0091	3·50	Cd ·0077	18·8
8	Sn ·0042	7·0	Sn ·0078	9·7	Ni ·0474	6·00	Cu ·0032	8·42	Ni ·0151	5·03	Cu ·0062	10·30	Cu ·0028	15·55
9	Pb ·0031		Ni ·0039		Al ·0215		Ni ·0014	5·60	Cu ·0133	16·60	Ni ·0057	11·40	Ni ·0022	10·00
10	Ni ·0027	5·4	Ag ·0026		Pt ·00015		Ag trace							
11	Fe ·0010													
12	Pt none													
	Average...	20·0	Average ...	7·52	Average ...	56·35	Average...	32·31	Average...	5·60	Average ...	17·14	Average...	70·95

that if two pieces of the same metal are connected with a galvanometer, their free ends immersed in an electrolyte, and heat applied to one of the junctions of the metal with the liquid, an electric current is produced. I have investigated this phenomenon in several researches (see *Phil. Mag.* Jan., 1857 ; *Proc.* Roy. Soc., 1871, Vol. XIX., p. 234 ; *ibid.*, 1878, Vol. XXVII., pp. 272 and 513 ; 1879, Vol. XXIX., p. 472 ; 1880, Vol. XXXI., p. 244 ; 1883, Vol. XXXVI., p. 50 ; and 1884, Vol. XXXVII., pp. 251-290), and have found that it is due to two chief causes, viz., 1st, to ordinary thermo-electric action in those cases where neither of the pieces of metal is at all corroded ; and, 2nd, to an alteration of ordinary voltaic potential in those where the metal is corroded. The most usual effect observed was, that in alkaline liquids the hot metal was often positive and in acid ones often negative ; the latter effect occurs with copper in an acidulated solution of cupric sulphate. In a large majority of cases the hot piece of metal became positive. The effect of heating both pieces was largely composed of the sum of the effects of heating each singly.

Theory of Voltaic Action and Source of Current.—Two distinct and different theories of the source of the voltaic current have long been entertained ; first, that of Volta and many Continental investigators, that the current is due to contact of dissimilar conductors of electricity ; and, second, that of Faraday and other English experimentalists, that it is due to chemical action. Neither of these views, however, is alone completely satisfactory or has been universally accepted, and the most general one held at present is that the current is due both to contact and to chemical action.

If we adopt the theory that the molecules of substances— those of chemically energetic bodies in particular—are in a state of ceaseless motion (that of frictionless bodies in a frictionless medium, such as the universal ether), and that when they chemically unite the amount of their motion is diminished, an efficient cause of chemical action and of voltaic current becomes much more clear.

According to this view, which has been gradually developed by the labours of many eminent investigators, neither contact nor chemical action is the primary cause of the current, but the essential cause is the stored up and ceaseless molecular energy of the corroded metal and of the corroding element of liquid with which it unites, whilst contact is only a static condition, and chemical action is the process or mode by which the molecular motions of those substances are more or less transformed into heat and current.

Both the heat and electric current produced by chemical corrosion of metals in electrolytes are recognised forms of energy, and as motion or energy cannot be created it must come from the original substances, and these must, after the action, have lost some of their *vis viva* and power of producing further heat or current.

SECTION E.

CHIEF FACTS OF ELECTRO-CHEMICAL ACTION.

Definition of Electrolysis.—Electrolysis, or electro-chemical action, is the decomposition of liquids by means of electric currents. The fundamental basis of electrolysis is the fact, *that when a current of electricity passes through a suitable liquid the liquid is decomposed.* The only essential conditions necessary are that the substance be a liquid, a definite chemical compound, and a conductor of electricity. Liquid alloys, such as that of sodium and potassium, tin and mercury, &c., are not decomposed by the current, and liquid compounds of two non-metallic elements, such as bisulphide of carbon, chloride of sulphur, &c., are also not usually capable of electrolysis. Electrolysis is commonly distinguished from ordinary chemical action by not taking place with non-conducting substances (pp. 51 and 87).

Distinction between Electrolytic and Voltaic Action.— Electrolysis, or electro-chemical action, is the converse of voltaic or chemico-electric action ; whilst electrolysis is chemical change produced by electric current, voltaic action consists of electric current produced by chemical change. Voltaic action is essentially a producer, and electrolysis usually a consumer, of molecular energy. In a voltaic cell substances are usually burned ; whilst in an electrolytic one they are usually unburned. In a voltaic cell elementary substances unite together, and their molecular energy assumes a free and active state ; in an electrolytic one, elementary substances are liberated at the poles, and absorb and render latent molecular energy. The one kind of action converts potential energy into active, and the other converts active energy into potential.

Connection between Electrolytic and Voltaic Action.—As in nearly every voltaic cell the current produced by the union of bodies at the positive plate decomposes the liquid at the negative one, and in nearly every electrolytic cell, voltaic action is produced by the substances set free at the electrodes, nearly every voltaic action produces electrolytic ones, and nearly every electrolytic action produces voltaic ones. According to these views pure voltaic action is essentially a case of chemical union, and pure electrolysis essentially consists of chemical separation.

Connection between Electrolytic, Chemical, and Voltaic Action.—The various phenomena of electrolysis are produced, not only by electric currents proceeding from an external source, but also by those originated in the electrolyte itself, and not only by currents flowing in circuits of measurable magnitude in that liquid, but also by others in circuits so small that they cannot be measured. In the case of an ordinary electrolysis vessel (or in a voltaic cell) the positive and negative surfaces are sufficiently far apart to enable us to perceive and distinguish the action at each ; but in that of " deposition by simple immersion," or the chemical precipitation of one metal by another, as when iron becomes coated with copper by simply immersing it in a solution of cupric sulphate, the positive and negative surfaces of each circuit are so extremely small, so exceedingly near together, and the circuits so very numerous, that they cannot be separately observed, and the entire immersed surface of the metal is covered with inseparable voltaic, chemical, and electrolytic actions.

The substances also set free by electrolysis do not always appear; the instant they are liberated they are subject to ordinary chemical action by contact with the liquid and the atmosphere. Thus when potassium is set free by electrolysis at the cathode from a solution of any of its salts, it is instantly oxidised into potash, or when oxygen is liberated at an anode in a solution of argentic nitrate, it at once combines with the silver of that salt. With rapidly reversed currents, the products do not always appear.

These facts show the intimate connection between electrolytic, chemical, and voltaic action, and the necessity of the student possessing a previous knowledge of chemistry and of voltaic electricity.

Arrangements for Producing Electrolysis.—Various combinations have been employed in which voltaic or other electric

currents produce chemical changes, and these have been classified as follows :—1. Electrolysis by simple contact of one metal
with one liquid.   2. By contact of one metal with two liquids.
3. By contact of two metals with one liquid.   4. By contact of
two metals with two liquids.   5. By a separate electric current.
And, 6. By a separate current and a series of electrolysis
vessels.   In Nos. 1, 2 and 3, the source of the electric current
is voltaic, and exists in the metals and liquids themselves,
whilst in 4 and 5 the current is generated separately by any
convenient voltaic or other method.

FIG. 21.   FIG. 22.
Simple Immersion Cell.   Two Liquids and One Metal Cell.

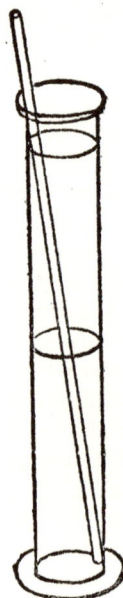

No. 1 arrangement is termed the "simple immersion process,"
and consists simply of immersing a piece of suitable metal in a
suitable liquid (Fig. 21).   The most familiar example of it is
the coating of iron with copper by simply dipping it into a
solution of blue vitriol.   In this process the voltaic currents are
generated by chemical action of the liquid upon the iron; they
are excessively minute, and produced in immense numbers at
points inconceivably small all over the immersed surface of the

metal, and re-enter all over that surface ; where they leave iron is dissolved, and where they enter copper or hydrogen is deposited. The method is extensively used for separating copper from solutions by means of scrap iron. By this method the electrolyte rapidly becomes impure.

No. 2 consists in either carefully placing a lighter liquid upon a heavier one in a tall narrow vessel, and standing a rod of metal vertically in the two strata (Fig. 22), or dividing a glass vessel into two parts by means of a vertical porous partition, placing the two liquids in the two divisions and immersing the two ends of the bent rod of metal in the two solutions (Fig. 24). The one end of the rod then dissolves and produces a voltaic current which re-enters the rod at its other end and deposits the metal.

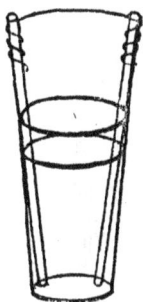

Fig. 23.
Two Metals and One Liquid Cell.

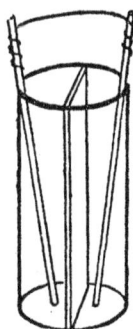

Fig. 24.
Two Metals and Two Liquids Cell.

By this contrivance the negative end of a piece of metal may be caused to receive an electrolytic deposit in a liquid which the metal itself is unable to decompose by simple immersion. Copper in dilute sulphuric acid, and in a saturated solution of cupric sulphate, is an example of this method.

No. 3 consists in bringing two metals into electrical contact at their upper ends, either direct or through a wire, and immersing their lower ends in a suitable liquid, or allowing the metals to touch each other in the solution (Fig. 23). A current then passes from the positive metal through the liquid into the negative one, producing deposition, and returns by the point of metallic contact; the positive metal also acts simultaneously by the " simple immersion process," and deposits metal upon itself. This contrivance, like the second one, enables a metal to receive a deposit in a liquid which it does not decompose by "simple immersion." Zinc in contact with copper, in a solution of cupric sulphate, is an example of this method. Under this arrangement may be classed the "two metal couples" of Gladstone and Tribe, in which the resistance to conduction is greatly diminished, and therefore the strength of current much increased, by making the voltaic circuits extremely small and numerous. This is effected by electrolytically depositing copper, silver, or platinum, in a porous, spongy state, upon the surface of zinc or magnesium, washing the plate so prepared, and immersing it in the liquid to be electrolysed.

No. 4 is termed the "single cell process," and consists of two liquids separated by a porous partition, the two metals being

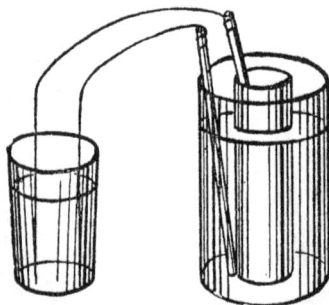

FIG. 25.—Separate Cell Process.

partly immersed, one in each liquid, and either in immediate contact or connected together at their upper ends by means of a wire (Fig. 24). This method also enables a metallic deposit to be produced upon a metal in a liquid which it does not itself decompose by simple immersion. In this arrangement, and in the second one, deposition by simple immersion is obviated by keeping the liquids apart by means of a porous partition; they, however, slowly diffuse through the partition, and the positive metal then becomes wasted by "simple immersion process."

No. 5 is the most convenient arrangement, and the most commonly used; it consists of a vessel containing the electrolyte and two electrodes, neither of which decomposes the solution by simple contact, the electrodes being connected by means of wires with a voltaic battery or other source of current. It is known as the "battery process" or "separate current process" (Fig. 25). By this method the strength of the current to produce electrolysis may be increased to any extent by means of additional voltaic cells, the most incorrodible metals may be employed as anodes, and by using a sufficiently powerful current even the alkali metals may be deposited. It was by this process that Davy first isolated potassium and sodium.

No. 6 consists merely of a series of such depositing vessels and electrodes as those in No. 5, with an undivided current passing through the whole of them (Fig. 26). It is now much used in copper refining.

Self-Deposition of Metals.—In addition to the cases in which a metal in contact with one liquid causes the deposition of the same metal upon itself in another liquid (*see* "No. 2," p. 74), instances have been observed by Raoult, Gladstone and Tribe,

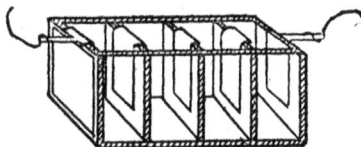

FIG. 26.—Series of Cells Process.

in which a metal deposits itself upon another less positive than itself, by a modified "simple immersion process." Raoult states that when two plates, one being of copper and the other of cadmium, are completely immersed in a solution of sulphate of cadmium deprived of air, and covered with a layer of oil, as long as they do not touch each other a very slight escape of hydrogen is observed on the cadmium plate, whilst the copper shows no visible change. When, however, the plates are caused to touch each other, cadmium begins at once to be deposited on the copper. Couples formed of gold-iron, gold-nickel, gold-antimony, gold-lead, gold-copper, gold-silver, immersed either in cold or hot, acid or neutral, solutions of salts of the more positive of the two metals, yielded, however, no deposit of that metal (*Comptes Rendus*, Vol. LXXV., p. 1,103; *Jour.* Chem. Soc., 2nd Series, Vol. XI., p. 464). Gladstone and Tribe also noticed that a copper zinc couple separated metallic zinc from a 1·5 per cent. solution of zinc sulphate (*ibid.*, p. 453).

Modes of Preparing Solutions for Electrolysis.—The special methods of preparing particular substances for making solutions differ in nearly every different case, and a description of them belongs to a work on chemistry. There are, however, two general ones for preparing the solutions, viz., the chemical and the electro-chemical process, the former being used for large operations and the latter for small ones. In the former the usual chemical processes of oxidation, solution in acids, crystallisation, &c., are employed; in the latter we take the particular solvent, usually a dilute acid, solution of potassic cyanide, &c., suspend in it a large anode of the particular metal to be deposited, and a smaller cathode, the latter being either in the solution itself or in a small porous cell filled with the liquid and placed in the bath, and pass a strong current until a sufficient quantity of the metal is dissolved, and the outer bulk of liquid yields a proper deposit with a current of suitable strength. The two processes, however, do not always yield liquids of exactly the same composition, because in the electrolytic one chemical changes occur at the cathode, and yield new products; for instance, with a solution of potassic cyanide, potassium is set free at the cathode, and is instantly oxidised by the water, and forms potash which dissolves in the liquid, and this potash gradually absorbs carbonic acid from the atmosphere and becomes carbonate of potash.

Nomenclature of Electrolysis.—The liquid undergoing electrolysis is termed an *electrolyte*. The immersed surfaces of metal or other conductor, by which the current enters and leaves the liquid, are called *electrodes*, or poles; the one by which it enters is called the *anode*, or positive pole, and that by which it leaves is the *cathode*, or negative pole. The substances into which the liquid is decomposed are called *ions*, those which separate at the anode being called *anions*, and those at the cathode *cations*. Anions are what are called electro-negative bodies, such as metalloids (fluorine, oxygen, chlorine, bromine, iodine, sulphur, phosphorus, &c.), acids, peroxides, &c.; and *cations* are electropositive ones, such as hydrogen, the metals, alkalies, and basic oxides. Hydrogen is the only known gaseous cation. As the conditions of positive and negative are not absolute, but only relative ones, the relatively negative constituents of the electrolyte are set free at the positive electrode, and the relatively positive ones at the negative electrode, and it sometimes happens that the same substance is separated at the anode in one liquid and at the cathode in another, according to the electric nature of the body with which it is united. Thus, sulphur, when united to a positive substance, such as a metal, as in sulphide of potassium, is separated at the anode, but when united with a more negative one than itself, such as oxygen, as in a solution of sulphurous anhydride, it is separated at the cathode; in a similar manner, iodine sometimes behaves as an anion and sometimes as a cation, as in the instances of potassic iodide and iodic acid. In the case of electrolysis of two liquids, separated by a porous partition, it is sometimes convenient to call the liquid containing the anode the *anolyte*, and that containing the cathode the *catholyte* (*see* p. 93).

Locality of Electrolysis.—The chemical changes directly produced by the current do not take place in the mass or body of the liquid, but at the immersed *surfaces* of the conductors by which the current enters and leaves the solution, and are strictly limited to the extremely thin layers of metal and liquid in immediate contact with each other.

Faraday proved that the decomposing action of an electric current upon an electrolyte is not necessarily limited to the contact surfaces of the solid or metallic conductor with the liquid, but may also occur at the mutual contact surfaces of two different electrolytes. Thus, by placing a concentrated solution of sulphate of magnesia of some depth below a deep stratum of distilled water, and passing an electric current from a platinum anode at the bottom of the magnesium solution upwards through the two liquids into a platinum cathode in the upper part of the water, magnesia was separated, not at the cathode, but at the surfaces of mutual contact of the two liquids. The water remained quite clear, no alkali was set free at the cathode, but plenty of acid was liberated at the anode. Daniell also subsequently, in a research on the "Electrolysis of Secondary Compounds" (*Phil. Trans. Roy. Soc.*, 1840, pp. 209-224), passed an upward electric current through solutions of various easily-reducible metallic salts into a supernatant dilute one of caustic potash, the two liquids being separated from each other by a thin horizontal film of bladder. Oxygen was separated at the upper, and the respective metals and oxides of metals at the lower surface of the bladder.

From a number of results obtained in a series of experiments on the "Influence of Voltaic Currents on the Diffusion of Liquids" (*Proc.* Roy. Soc., Vol. XXXII., 1881, pp. 83 and 84), in which the effect of a vertical electric current passing through the horizontal surfaces of mutual contact of dissimilar electrolytes in a large number of instances was examined, the author concluded that "ions are probably liberated at every surface of junction of electrolytes, of sufficiently different composition, through which the current passes," and "that every inequality of composition or of internal structure of the liquid in the path of current, must also act to some extent as an electrode."

Distribution of Current in an Electrolyte.—With an electrolyte of perfectly homogeneous composition and temperature,

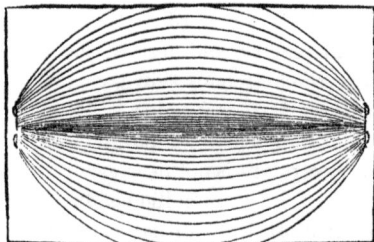

FIG. 27.—Equipotential Lines.

and two narrow vertical electrodes placed symmetrically in it, the current from the anode spreads out in curves of equipotential lines, not unlike those of magnetism diverging from the poles of a magnet, and converge in similar lines to the cathode, the densest portion of the current and the greatest number of lines being in the central axis of the liquid, and joining the centres of the electrodes (Fig. 27). Its distribution has been experimentally examined by Tribe, who suspended small pieces of sheet metal in different parts of the liquid between the electrodes, and ascertained it by the amounts of electrolytic action produced upon them by the same current during the same period of time (*Proc.* Roy. Soc., Vol. XXXI., p. 320, Vol. XXXII., p. 435; *Phil. Mag.*, June, 1881, p. 446; October,

1881, p. 299; June, 1883, p. 391; August, 1883, p. 90; and October, 1883, p. 269).

Conduction in Electrolytes without Electrolysis?—The question whether electrolytes conduct in some minute degree without undergoing decomposition has been an unsettled one during many years. If they conducted in this manner freely the results would be serious in electro-metallurgical processes on a large scale, because there would be great waste of current. That each electrolyte requires a certain electromotive force to decompose it is true, and that it transmits by some means even the feeblest current is also true. "A single Daniell cell connected with platinum electrodes in sulphuric acid produces only polarisation, no visible decomposition, the voltameter acting as a condenser of immense capacity" (*Jour.* Chem. Soc., 2nd Series, Vol. X., p. 463). It is well known that an electromotive force of about 1·47 volts is necessary in order to electrolytically decompose water; nevertheless, much below this difference of potential a current passes, and is easily detected by means of a galvanometer. Different views are held as to the mode of transmission of this current. One is that it is transmitted by convection of the liquid particles, like that which occurs when an electrically-charged body discharges itself by attraction and repulsion of particles of dust in the air; in this case the liquid is not decomposed. Another is that within these extremely minute limits the solution conducts like a metal; in this case also the liquid is not decomposed. A third is that at first the liquid is decomposed, but that the ions are not liberated but are occluded by the electrodes, and by their subsequent very gradual diffusion and reunion produce a very feeble current. And a fourth is that it is due to air dissolved in the liquid. There are also other views on the subject. Melted fluoride of silver *appears* to transmit a very much larger current than agrees with the amount of visible electrolysis.

Alternate-Current Electrolysis.—Various investigators have observed that when an electric current, the direction of which is continually and rapidly being reversed, passes through an electrolyte, products of electrolysis sometimes appear at the electrodes, whilst at other times they do not (see *The Electrician*, Vol. XXI., p. 403; *Nature*, Vol. XXXVIII., p. 555; *Phil. Mag.*, June, 1853, p. 392).

G

The explanation of this is simple, and the liberation of the products of electrolysis in such a case depends upon two conditions: first, the degree of frequency of reversal of the current; and second, upon the density of the current at the electrodes. If the reversals are sufficiently slow, or the quantity of ions separated per unit of surface is sufficiently large, there is set free at each electrode a mixture of positive and negative substances. It is manifest that if the reversals are very slow, the products of the portion of current from A to B electrode will have time to separate and get away from the electrodes before those of the succeeding portion of current from B to A can be liberated and recombine with them. It is also clear that with a given rate of reversal, if the density of the current is sufficiently great, the quantity of the products of the current from A to B will be so large that they will be produced so fast as to push each other away from the electrodes more rapidly, and thus get away from the electrodes before those of the succeeding current from B to A can be liberated and recombine with them. In each of these cases, also, the products will have lost their nascent state to a greater or less degree before those of the opposite kinds can come into contact with them, and therefore will have lost to a corresponding extent their power of spontaneously recombining.

Transport of Ions in Electrolysis.—According to the results of experiments made by Hittorf and F. Kohlrausch, "in very dilute solutions of various salts, of strengths proportional to the chemical equivalents of the salts, the specific conductivities of the solutions are all of the same order of magnitude. In very dilute solutions, acted upon by a current of given electro-motive force, each of the ions moves through the solution with a fixed velocity dependent upon its own chemical action, and independent of the velocity of the other ion, *i.e.*, the migrations of the different ions are independent of each other. The following is the order of velocity of cations, the first named being the fastest:—Hydrogen, potassium, ammonium, barium, copper, strontium, magnesium, zinc, lithium; and of anions, hydroxyl, iodine, bromine, cyanogen, chlorine, NO^3, ClO^3, and the halogen of acetic acid." The quickest is "about four inches per hour" (see *The Electrician*, Vol. XXI., pp. 466 and 622).

According to O. Lodge ("Modern Views of Electricity," 1889, p. 87), "The following are the rates at which atoms of various kinds can make their way through nearly pure water when urged by a slope of potential of one volt per lineal centimetre":—

	Centimetre per hour.		Centimetre per hour.
Hydrogen	1·08	Silver	·166
Potassium	·205	Chlorine	·213
Sodium	·126	Iodine	·216
Lithium	·094	NO³	·174

Electrolytic Osmose and Diffusion of Liquids.—In the year 1807, Reuss of Moscow discovered electric osmose; and in 1817, Porrett extended the discovery and modified the experiment, and showed that if an electrolyte is divided into two portions by means of a porous partition, and an electric current is passed from one portion to the other, the liquid flows slowly through the partition in the direction of the current, and increases in bulk on the negative side. In a research on the same subject (*Proc.* Roy. Soc., Vol. XXXI., 1880, p. 253) I have shown that the direction of flow is occasionally affected by the nature of the liquid, and that in a saturated alcoholic solution of bromide of barium it was opposite to that of the current. This was the only exceptional case in a series of sixty-eight different liquids of varied composition and strength. For additional experiments on the influence of electric currents on the diffusion of electrolytes, see *Proc.* Roy. Soc., Vol. XXXII., 1881, pp. 56-85.

Influence of Liquid Diffusion on Electrolysis.—Diffusion exercises very great influence upon electrolysis; if there was no diffusion or motion of the liquid, there could be no continued electro-deposition of metal, because the solution near the cathode would become exhausted of that substance, and that near the anode would be wholly deprived of free acid. The rate of electrolysis is largely limited by that of diffusion; if the electro-deposition of metal proceeds faster than the metallic salt diffuses from the mass of the liquid to the surface of the cathode, the current begins to decompose the other compounds present, usually the saline impurities and the water; if, also, it proceeds faster than acid can diffuse to the anode, the latter becomes coated with oxide, or faster than water can diffuse to

G 2

it, it becomes encrusted with salt (*see* p. 88). When the speed of diffusion is deficient in relation to that of electrolysis, some of the products of electrolysis are themselves liable to be electrolysed. With a viscous liquid the quality of the deposited metal soon deteriorates during electrolysis, because of deficiency of supply of metallic salt to the cathode; such a solution can only yield reguline metal very slowly.

Long has observed that in almost every case the best conducting aqueous saline electrolytes are those composed of salts which have the fastest rate of diffusion, and those are usually the ones which have the largest molecular volume, and which absorb the most heat in dissolving. He also arrived at the conclusion that the "rate of diffusion of a salt is proportional

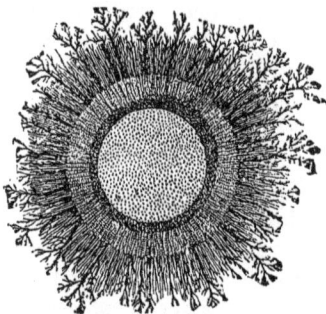

Fig. 28.—Decharme's Experiments.

to the sum of the velocities with which its component atoms move during electrolysis" (*Phil. Mag.*, 1880, Vol. IX., p. 425).

Directly a dense current enters the cathode in a sulphate of copper solution the layer of liquid next the cathode is deprived of nearly all its metal, and is thereby not only rendered specifically lighter, but also less viscous and more diffusible, and tends to pass away from the cathode faster than the denser liquid tends to approach. This is rendered manifest by the following experiments made by M. Decharme, who, however, gives a different explanation of the phenomenon. If we place a small disc of copper upon a film of solution of argentic nitrate covering a clean plate of glass, the liquid at once draws back from the edge of the disc as if it was strongly repelled, and in less than a minute a nearly dry annular space is produced all round the disc. In a few minutes the liquid gradually returns towards the disc. The first portion of silver separated is brown, and looks like imperfectly reduced amorphous oxide, in the form of an annular layer; succeeding this is a zone of white semi-crystalline spongy silver. Soon after this rapid action is

FIG. 29.—Decharme's Experiments.

over, and the silver solution has returned to the disc, the arborisations or tree-like formations of definite crystals of silver begin to form at the outer edge of the white deposit (see Fig. 28). If a disc of zinc was placed upon a film of pure water of the same thickness it produced around it merely the ordinary ascending meniscus. M. Decharme attributes the centrifugal motion of

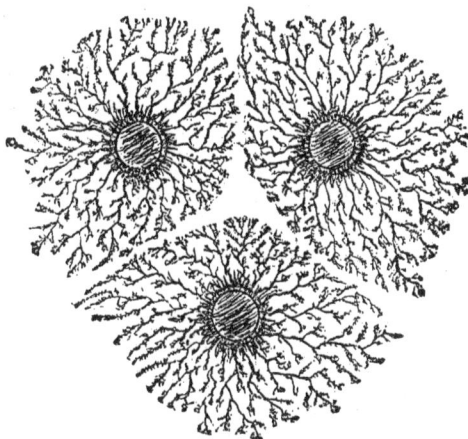

FIG. 30.—Decharme's Experiments.

the silver solution to "electric repulsion," but to me it appears to be due to the inner ring of liquid being deprived of its silver, and thereby rendered more readily diffusible. Figs. 29 and 30 are illustrations of the "repulsion" of silver arborisations (see La Lumière Électrique, 1887).

Polarisation of Electrodes, Counter Electromotive Force.— Immediately an electric current passes through an electrolyte products of electrolysis begin to be liberated, and accumulate at the surfaces of the electrodes, and directly this occurs voltaic action is produced, because the two electrodes are no longer alike in composition and no longer in contact with layers of liquid of the same nature. This dissimilarity of conditions excites a counter electromotive force, tending to stop the original current, and to produce a voltaic one in the opposite direction. The substances which produce this effect differ in almost every different case. They usually consist of thin films of either solids, liquids or gases adhering to the surfaces of the electrodes, and in many cases may be largely or entirely removed by rapid stirring, but they sometimes consist of gases absorbed by the electrodes. With a cathode of sheet platinum in distilled water I have found the resistance and counter electromotive force, due to the gas absorbed by the platinum, and not removable by rubbing, sometimes amount to as much as 18 per cent. of the total electromotive force of the current. (*See* p. 34.)

Unequal Electrolytic Action at Electrodes.—By the electrolysis of a solution of a metallic salt with vertical corrodible electrodes, the liquid about the anode usually becomes more saturated with metallic salt, and being rendered specifically heavier descends and forms a layer at the bottom of the vessel, whilst that about the cathode becomes deprived of metal, acquires less specific gravity, ascends and spreads over the surface of the electrolyte. In consequence of these differences of specific gravity and chemical composition of the upper and lower parts of the electrolyte, the distribution and direction of the electric current in it are affected. At the commencement, whilst the liquid is uniform in composition the current is uniformly distributed, and equal strengths of it pass horizontally through equal cross-sections of the liquid in every part ; but if the current is sufficiently strong, and produces the above inequalities of composition faster than diffusion or other motion corrects them, after a time it passes unequally, the greater portion of it going in an oblique direction from the upper part of the anode to the lower part of the cathode. In consequence of this the upper part of the anode becomes the most corroded, and the lower part of the cathode receives the greatest amount of deposit. These changes in composition of the liquid also give rise to local electric currents quite distinct from the general one ; for instance, as the upper part of each electrode is in contact with a more acid liquid, and the lower part with liquid containing more metallic salt, the upper part is continually being corroded and generating an electric current, which continually re-enters the lower part and deposits metal, and thus an anode sometimes actually increases in thickness at its lower end instead of becoming thinner, and is gradually cut off at the surface of the liquid and falls to the bottom. With horizontal electrodes and the anode above, these effects do not occur.

Influence of Chemical Composition of the Liquid on Electrolysis.—The chemical nature of the liquid is a matter of fundamental importance ; it determines whether electrolysis takes place at all, and the kind of effects it produces. Some fused salts, such as melted boracic acid, sulphide of arsenic, iodide of sulphur, and sulphide of phosphorus, scarcely conduct or suffer electrolysis at all. Some liquid compounds behave similarly; for instance, bisulphide of carbon, the chlorides of sulphur, phosphorus and arsenic, tetrachloride of tin, pentachloride of antimony, the chlorides of titanium and silicon, &c. Even some aqueous solutions of chemical compounds offer considerable resistance to conduction and electrolysis ; amongst these are boracic acid, cyanide of mercury, ammonia, sugar, the various fatty acids, &c. Even the addition merely of a trace of an extra ingredient to the electrolyte has, in some cases, a great effect ; thus the addition of a minute quantity of bisulphide of carbon to the ordinary silver-plating solution changes the deposit from dull white soft metal to hard and brilliant silver. The influence of the solvent is also very great; in some experiments " On the Electrolysis of Alcoholic and Ethereal Solutions of Metallic Salts" (*Proc.* Birm. Phil. Soc., 1887, Vol. V., Part II., p. 371) I have shown that such liquids are much less easily electrolysed than aqueous solutions of the same salts. I have also observed that whilst aqueous acidified solutions of teriodide and terchloride of antimony are easily electro-

lysed, solutions of the same substances in bisulphide of carbon
scarcely conduct at all. W. Hampe states (*Jour.* Chem. Soc.,
p. 211, March, 1888) that the following solutions are good
electrolytes :—Cupric chloride in concentrated alcohol; alcoholic
solutions of the iodides and bromides of zinc and cadmium ;
chloride and bromide of zinc dissolved in ether.

The influence of the chemical nature of the liquid is so
general that with every different electrolyte the phenomena and
products of electrolysis are more or less changed, and as the
number of possible electrolytes is extremely large, the electro-
chemistry of individual substances would fill a volume. A large
number of chemical compounds and mixtures have never yet
been subjected to the influence of the current. For information respecting the electrolysis of individual substances, and the
behaviour of those which have been electrolysed, the reader is
referred to a small book on "Electro-Chemistry" (1885), written
by the author of this work and published at the office of *The
Electrician ;* also to "D'Electrochimie," by D. Tommasi.

Influence of Water, Free Acid, &c.—In the case of aqueous
electrolytes (and these are usually the kind employed), if the water
is in large excess the liquid is liable to offer greater conduction
resistance and become more heated, and if the current is strong
some of the water is apt to be decomposed in consequence of
the relative deficiency of the dissolved substance ; and if the
water is deficient in amount the diffusive power of the liquid
is weakened, the liquid about the electrodes becomes exhausted
of material, and the products of electrolysis accumulate there.
The anode also, if a corrodible one, becomes covered with undissolved salts ; in cases where evaporation occurs freely, as
with heated liquids, the latter effect is very apt to occur. With
an acidulated aqueous electrolyte, if the acid is in large excess,
it is apt to be decomposed by the current, and the deposit upon
the cathode, if of an easily corroded metal, is deficient in
quantity, and the amount of salt in solution increases ; and if
the acid is deficient in quantity, the anode, if a corrodible one,
is liable to become coated with oxide, which impedes the
current ; other effects are also apt to occur according to the
nature of the electrodes and of the electrolyte. With a perfectly neutral solution of cupric sulphate, the deposited copper

is inferior in quality, and appears to contain cupreous oxide, because it has a *dull* pink colour and loses weight when heated in hydrogen. A number of metals are more readily separated from a non-acidified than from an acidified solution, because various acids are more easily decomposed than water.

Influence of Temperature on Electrolysis.—As the chief amount of conduction resistance is in the electrolyte and not in the electrodes, and by rise of temperature the resistance of liquids is more decreased than that of metals is increased, the balance of effect of rise of temperature is to diminish the total resistance and facilitate electrolysis. It also increases the solubility of the dissolved salt and the diffusive power of the ingredients of the liquid ; it, however, at the same time largely increases the corrosive power of the solution and diminishes the proportion of metal deposited upon the cathode to that dissolved at the anode (*see* page 123). If an electrolyte is constantly heated, the proportion of free acid in it quickly decreases, and that of dissolved metallic salt equally increases in consequence of the more rapid chemical corrosion ; the liquid also rapidly becomes saturated with metallic salt, partly from this cause, but chiefly from evaporation of water. According to Dr. Hammerl, heating the ordinary coppering solution to boiling causes the deposited copper to be almost completely changed into cupreous oxide (*Nature*, Vol. XXIX., p. 227).

Thermal Phenomena attending Electrolysis.—Not only do alterations of temperature largely affect chemical action and electrolysis, but, conversely, in consequence of chemical actions and of the passage of the current from one substance to another, changes of temperature occur at each electrode, and in cases where two electrolytes are employed also at their surfaces of mutual contact. By making a thin platinum bottle (Fig. 31) the cathode in dilute sulphuric acid or dilute nitric acid, and fixing to it a narrow open glass tube with a drop of water in it as an index, on suddenly passing a current, the air in the bottle at once expands, showing a production of heat at the electrode ("Evidence Respecting the Reality of 'Transfer Resistance' at Electrodes": *Proc. Birm. Phil. Soc.*, 1885-6, Vol. V., p. 27 ; *Phil. Mag.*, Feb., 1886).

As the decomposition of a compound absorbs the same quantity of heat as that which was evolved by its formation, the loss
of thermal energy at the cathode in an electrolysis cell equals
the gain of that at the anode, provided a similar quantity of
the same metal is deposited in a like physical state as that
which was dissolved at the anode; the final thermal result is
also independent of the kind of salt formed, or of the changes it
undergoes as an electrolyte; all this agrees with the principle
of conservation of energy (*see* p. 38). For instance, at the surface of an anode composed of copper in an acidulated solution
of cupric sulphate, heat is evolved by the oxidation of the
copper, and by the union of the oxide with the acid and the

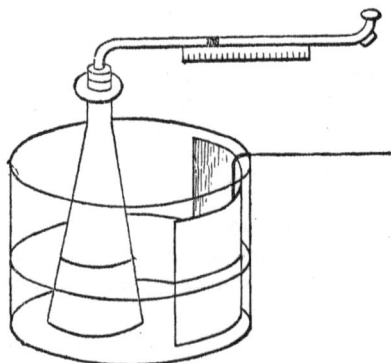

Fɪɢ. 31.—Platinum Bottle Experiment.

dissolving of the salt in the water; at the same time heat
is absorbed in equal amount at the cathode by the exactly converse process of separation of copper. In addition to these a
small amount of heat is evolved by chemical corrosion at each
electrode, and in the mass of the liquid by conduction resistance. By passing a powerful electric current through a small
glass tube uniting two large volumes of an electrolyte I have
caused the liquid to boil by the heat of conduction resistance
(" Influence of Voltaic Currents on Diffusion of Liquids," *Proc.*
Roy. Soc., 1881, No. 213, pp. 76-82). With platinum electrodes
in dilute sulphuric acid heat is absorbed at the anode by liberation of oxygen.

Electrolysis of Fused Compounds.—Amongst those compounds which have been electrolysed are the following:—Oxide of bismuth, oxide of lead, the hydrates of potassium and sodium, nitrate of silver, nitrate of ammonium, cyanide of potassium, cyanide of caesium, fluoride of tin, cryolite, a mixture of potassic fluoride and oxide of cerium, the chlorides of silver, lead, tin, zinc, magnesium, cerium, didymium, barium, lithium, sodium, potassium, and rubidium, double chloride of aluminium and sodium, mixed chlorides of calcium and strontium, chlorate of potash, iodide of potassium, carbonate of sodium, borax, pyrophosphate of sodium, tersulphide of antimony, &c. I have electrolysed the melted fluorides of silver, copper, lead, manganese, uranium, lithium, potassium, ammonium, hydro-potassic fluoride, borofluoride of potassium, silicofluoride of potassium, mixed fluorides of magnesium and calcium, molybdic acid, silicate of soda, tungstate of soda, bichromate of potash, a mixture of the carbonates of potash and soda, &c. (see " Electro-Chemistry," published at the office of *The Electrician*).

According to W. Hampe (*Jour.* Chem. Soc., p. 211, March, 1888), the following salts are good electrolytes when in a melted state :—All the haloid salts of barium, lithium, sodium, potassium, rubidium, and caesium, the chlorides of barium, strontium, calcium, magnesium, zinc, cadmium, beryllium, lanthanum, didymium, cerium, tantalum, thorium, and lead ; the bromides of strontium, calcium, beryllium, magnesium, zinc, cadmium and lead ; thallous, gallous, stannous, and cupreous chlorides ; the iodides of zinc, cadmium, and lead ; melted tersalts of antimony. Fused aluminium chloride and bromide do not conduct.

Gladstone and Tribe have shown that if a bent strip of metal has its two ends immersed in a fused salt of the same metal, the liquid being of different temperatures at the two parts where the metal dips into it, the hotter end of the metal dissolves and the less heated part receives a metallic deposit ; copper in fused chloride of copper is an example (*Jour.* Chem. Soc., Vol. XI., p. 868, 1881). This is an instance of electro-deposition by "arrangement No. 2 " (see p. 75). I have also made and published a large number of somewhat similar experiments relating to the electrical effects of unequally heated metals in aqueous electrolytes, and have found that whilst in

many cases the hot metal was positive, in many others it was
negative; the former happened chiefly in alkaline liquids, and
the latter in acid ones (*see* p. 70).

Influence of the Kind of Electrodes.—The effects of the
current are modified to some extent by the nature of the electrodes. This is largely due to ordinary physical and chemical
action, and depends not only upon the nature of the electrodes
but also upon that of the substances in contact with or
deposited upon them. If hydrogen is deposited upon a cathode
of palladium it is absorbed and the palladium expands, but if
it is deposited upon copper there is very little effect. If the
cathode is composed of mercury it absorbs nearly all metals
deposited upon it; it absorbs pure antimony, but not the black
explosive variety. By employing a mercury cathode various
investigators have been enabled to separate some of the most
highly oxidisable metals, such as chromium, &c., in the form of
an amalgam, and then obtained the metal by distilling away
the mercury.

Metallic Sulphides, &c., as Electrodes.—In addition to
metals and alloys any insoluble chemical compound which conducts electricity may be employed as an electrode; in consequence, however, of the inferior conductivity of artificial compounds and native minerals (*see* p. 33), and the usual difficulty
of obtaining them in the form of rods or plates, they are rarely
employed.

Amongst these substances used as anodes are peroxide of
lead, which has the property of resisting chlorine, also plates of
matte or mineral regulus formed by melting the mixed sulphides of iron and copper (*see* pp. 33, 229, 230). When the
latter are used as anodes, the metal in them usually dissolves,
and the sulphur separates in the elementary state.

According to Badia (*La Lumière Électrique*, Vol. XIV., 1884),
" if we use as an anode mineral cupric sulphide (not cupreous sulphide) in a solution of cupric sulphate, it takes less than one
volt of electromotive force to electrolyse the solution, and as
fast as copper is deposited at the cathode an equal amount is
dissolved at the anode, until nothing but sulphur is left. It is
the same if a solution of cupric nitrate is employed; the solid

sulphide is decomposed, all the copper is dissolved out of it, no sulphur is oxidised and no sulphuric acid is formed, and the whole of the copper may be deposited out of the solution by electrolysis.

" With an anode of sulphide of iron in dilute sulphuric acid, no sulphide of hydrogen is evolved as long as the current passes, and with the same anode in a solution of cupric sulphate, copper begins to separate at the cathode and sulphur at the anode, when the difference of potentials is much less than one volt, and by continuing the current long enough, the solution is deprived of all its copper, and then consists of dilute sulphuric acid, ferrous sulphate, and basic persulphate of iron ($Fe^2O^32SO^3$). Copper of good quality, but rapidly diminishing in quantity, deposits as long as the solution contains ·1 gramme of copper per 100 cubic centimetres of liquid; it then becomes black, and hydrogen is evolved. As during this action the strength of the current does not decrease, whilst the rate of deposition of copper diminishes, the sulphur which separates, being loose and porous, does not increase the resistance; and the current does some other work besides that of depositing copper, it deoxidises ferric salt at the cathode.

" With an anode of sulphide of iron and a cathode of cupric sulphide, in the two compartments of a cell filled with a solution of cupric sulphate and divided by a porous partition, on passing a current having an electromotive force of one volt, bright copper is separated upon the cathode, and sulphur and undecomposed sulphide at the anode; the catholyte becomes weaker in cupric sulphate, and acquires a little ferrous sulphate, and the anolyte loses all its copper salt, and acquires free sulphuric acid, ferrous sulphate, and basic persulphate of iron. The amount of copper deposited is nearly equivalent to the total amount of current passed; but by adding a few drops of a solution of ferric sulphate to the catholyte, the weight of copper separated is no longer equivalent to the quantity of current, but is deficient, although the deposited copper continues pure and bright, and the ferric salt is reduced to ferrous sulphate. As soon, however, as the persalt is most of it reduced, the copper begins to deteriorate in quality, and hydrogen is set free.

" With an anode of sulphide of iron, and an acidulated solution of mixed ferrous and ferric sulphates in the divided cell, and a current of less electromotive force than one volt, ferrous salt is oxidised at the anode and ferric salt deoxidised at the cathode, and hydrogen is set free.

" With the same mixed liquid and an anode of cupric sulphide in the divided cell, cupric sulphate is formed, and a little of the ferrous salt peroxidised at the anode, the electric action being divided, and hydrogen is evolved at the cathode.

" As copper mattes, formed by the fusion of cupreous pyrites, *i.e.*, native sulphides of iron and copper, are fairly good conductors of electricity, cast plates of them may be used as electrodes. Poor mattes consist chiefly of ferric sulphide (Fe^2S^3) and cupric sulphide (CuS), whilst rich ones contain, in addition, ferrous sulphide (FeS), some cuprous sulphide (Cu^2S), copper oxides, and sometimes even metallic copper in fine filaments. All such mattes, when used as anodes, behave as the results of the above experiments indicate, *i.e.*, in a solution of cupric sulphate, with a copper cathode, they are decomposed, and yield a deposit of copper by passing through them a current having an electromotive force of not more than one volt, and produce in the solution free sulphuric acid, ferrous sulphate, and basic persulphate of iron."

The results of these experiments show that whilst the presence of ferric salt in the solution entirely prevents the separation of hydrogen and deterioration of the quality of the copper when the density of the current is large in proportion to the amount of dissolved copper, it is attended by loss of energy expended in reducing that salt to the ferrous state; it is highly desirable, therefore, to keep all persalt of iron out of the liquid.

The results also show that with anodes composed of sulphides of copper and iron, containing three equivalents of copper to two of iron, in sulphate of copper solution, the current passes equally between the copper and the iron, because the resistance of the two sulphides are about equal. The resistance of the mixed sulphides (excluding that of Cu^2S) is much nearer that of metals than that of the electrolyte. The decomposition of these sulphides consumes less energy than that required by a solution of cupric sulphate (*Scientific American Supplement,* 1885, No. 478).

Influence of Solubility of Anodes.—Both anodes and cathodes which are not readily corroded by chemical action resist the passage of the current at their surfaces of mutual contact—for instance, platinum in various electrolytes—whilst easily corroded ones allow the current to pass readily. If the anode is not chemically corroded by the liquid, nor by the products of electrolysis, the current is not only greatly retarded, but the electro-chemical products at its surface are different ; for example, with an anode of platinum or iron in a solution of potassic cyanide, considerable electromotive force is required to produce electrolysis, gas is evolved at the anode, and the latter is not corroded ; but if the anode is composed of silver, the current passes easily, no gas is set free, and the anode dissolves rapidly. Whether the whole of this extra energy required with a platinum anode is absorbed by the liberated elementary anion, or a portion is expended in overcoming "transfer resistance," is an interesting question. In some cases it takes more than four times the electromotive force, or more than four times the period of time, to deposit the same amount of copper with a platinum anode as with a copper one. With an anode composed of an alloy, the most easily corroded metal, or the one which evolves most heat in corroding, is attacked the first, and if it happens to form an insoluble compound, that compound either adheres to the anode and impedes the current or falls to the bottom.

The following tables exhibit the different amounts of retarding influence, expressed in ohms, offered to a current from 50 Pabst elements in passing into and out of different liquids by means of vertical and parallel sheet electrodes of different metals $1 \cdot 0 \times 1 \cdot 0$ centimetre, and $5 \cdot 0$ centimetres apart, the liquid near the electrodes being rapidly stirred during each experiment in order to diminish or prevent polarisation. The substances to be electrolysed were taken in the proportions of their chemically equivalent weights, and each substance was dissolved in 250 cubic centimetres of distilled water in a circular glass cup $9 \cdot 5$ centimetres wide, with the liquid in it $3 \cdot 5$ centimetres deep.

Retardation at Electrodes.

Nitric Acid, HNO^3.　　Hydrofluoric Acid, HF.

·63 gramme. Solution at 9°C.					·20 gramme. Solution at 8°C.		
	Anode.	Cathode.	Total.			Anode.	Cathode.	Total.
Zn	5	4	9	Zn	53	58	111
Cd	6	5	11	Cd	65	96	161
Pb	7	65	72	Pb	47	94	141
Sn	7	9	16	Sn	83	92	175
Fe	5	17	22	Fe	69	79	148
Ni	143	49	192	Ni	174	80	254
Cu	14	59	73	Cu	72	126	198
Ag	31	67	98	Ag	112	132	244
Au	130	83	213	Au	187	215	402
Pd	116	94	210	Pd	148	216	364
Pt	123	79	202	Pt	160	194	354
	587	531	1118		1170	1382	2552
Average	53	48		Average	106	126	

Hydrochloric Acid, HCl.　　Sulphuric Acid, H^2SO^4.

·365 gramme. Solution at 9°C.					·49 gramme. Solution at 9°C.		
	Anode.	Cathode.	Total.			Anode.	Cathode.	Total.
Zn	9	26	35	Zn	9	30	39
Cd	10	58	68	Cd	10	60	70
Pb	9	75	84	Pb	207	96	303
Sn	13	82	95	Sn	12	71	83
Fe	14	45	59	Fe	15	28	43
Ni	24	72	96	Ni	157	55	212
Cu	21	86	107	Cu	15	83	98
Ag	36	91	127	Ag	44	108	152
Au	138	121	259	Au	145	114	259
Pd	133	103	236	Pd	112	123	235
Pt	124	98	222	Pt	109	114	223
	531	857	1388		835	882	1717
Average	48	78		Average	76	80	

Average "Retardation."

	With Zinc.		With Platinum.	
	Anode.	Cathode.	Anode.	Cathode.
In 6 acid liquids	15·	29·	97·	117·
,, 29 neutral liquids..	24·5	59·8	128·	153·
,, 18 alkaline liquids.	143·2	68·3	133·5	159·3

The extra retarding influence with anodes of zinc in alkaline liquids was probably due to adhering coatings of oxide. Why the hindrance to the current should usually be greater at the cathode than at the anode is an interesting inquiry ("Relations of 'Transfer-resistance' to the Molecular Weight, &c., of Electrolytes," *Proc.* Birm. Phil. Soc., 1887, Vol. V., Part II.; *see* also references to Papers on "Transfer-resistance," on p. 34).

The following are some unpublished measurements of "transfer-resistance" at electrodes in several of the ordinary electro-depositing solutions :—

(1) *Acidified Sulphate of Copper Solution at 7°C.*—Copper electrodes 2 millimetres wide, 1 centimetre long, 2 centimetres apart. Liquid stirred. Resistance of the liquid 18 ohms. Strength of current, ·0364 ampere. Reguline deposit. Anode black—

Total.	Anode.	Cathode.
19 ohms	8 ohms	11 ohms

With strength of current = ·0068 ampere. Reguline deposit. Anode black—

27 ohms	10 ohms	17 ohms

(2) *Sulphate of Nickel and Ammonium Solution at 13°C.*— Nickel sheet electrodes, not varnished on their backs, 1·0 centimetre wide, 1·0 centimetre long, and 2·0 centimetres apart. Liquid stirred. Resistance of the liquid, 28 ohms. Strength of current, ·01 ampere—

Total.	Anode.	Cathode.
202 ohms	133 ohms	69 ohms

With strength of current = ·005 ampere—

278 ohms	180 ohms	98 ohms

(3) *Cyanide of Silver and Potassium Solution at 13°C.*— Silver electrodes, not varnished on the backs; 1 centimetre long, 1 centimetre wide, and 2 centimetres apart. Liquid stirred. Resistance of the liquid, 14 ohms. Strength of current, ·01 ampere—

Total.	Anode.	Cathode.
20 ohms	14 ohms	6 ohms

With strength of current = ·005 ampere—

22 ohms	15 ohms	7 ohms

H

Insoluble Coatings on Electrodes.—In many cases, with aqueous solutions, the anode does not dissolve, but becomes coated with an oxide, fluoride, chloride, cyanide, or other insoluble salt, usually caused by chemical union of the metal with a constituent of the liquid. In this manner a silver anode in dilute hydrochloric acid becomes coated with chloride, or in a solution of argento-cyanide of potassium it becomes covered with argentic cyanide ; lead in dilute hydrofluoric acid becomes coated with fluoride, aluminium in many liquids becomes covered with an invisible thin film of oxide, &c., which greatly hinders the current. In some cases the insoluble coating is produced, not by corrosion of the anode, but by the liberated anion acting chemically upon the liquid. In this way liberated oxygen produces films of peroxide upon a platinum anode in solutions of the nitrates of bismuth, silver, and manganese; in those of the acetates of lead and manganese; in some solutions of nickel and cobalt ; and in alkaline ones of lead. Some of these films, when they are exceedingly thin, exhibit magnificent colours, and are therefore employed for ornamenting metallic articles, the process being termed " metallo-chromy." These various coatings greatly retard the current in nearly all cases, because they are usually very bad conductors ; they also affect the products of electrolysis at the anode.

Upon the cathode insoluble coatings occur very much less frequently than upon the anode ; they happen usually when a suboxide of a highly oxidisable metal is separated, as with some solutions of salts of manganese and cobalt. A film of hydrogen also sometimes adheres to the cathode. Each of these coatings diminishes the current.

Influence of Density of Current.—Increased density of current is attended by increased heat of conduction resistance in the liquid ; and if the increase of density is great in a copper depositing solution, gas is evolved at the anode and a deficiency of copper is dissolved (*see* also pp. 110 and 111).

Quincke has shown that the force tending to separate the elements of an electrolyte is proportional to the strength of the current per unit of sectional area of the liquid (*Jour.* Chem. Soc., Vol. X., p. 208). The degree of density of current at the surfaces of the electrodes is an important circumstance ; variation of it has often very great effect, both upon the chemical composition and the physical structure of the ions. It appears to influence the properties of oxygen and chlorine gases evolved at the anode. I observed, whilst electrolysing pure dilute hydrofluoric acid containing about 40 per cent. of water with platinum electrodes, that if the current was strong ozone was evolved, but if it was weak odourless oxygen was set free. It was chiefly by employing great density of current that Sir H. Davy succeeded in first isolating potassium and Bunsen separated chromium, and that other difficultly reducible metals have been electrolytically obtained (see also "Divided Electrolysis," p. 116).

It is not by misdirected strength of current, however great, that difficultly reducible elements are separated. From a weak solution of a potassium salt even the strongest current will not enable us to obtain the metal; but by using a cathode of mercury of small surface, and thus protecting the deposited metal from the chemical action of the liquid, even potassium has been isolated with the aid of a feeble current. It was by using an extremely cold electrolyte and taking certain other precautions that Moissan electrolytically isolated fluorine. Electrolysis, therefore, is no exception to the general truth—that the secret of success in all things is intelligently directed energy.

Physical Structure of Electro-Deposited Metals.—The physical structure of every electro-deposited metal is profoundly affected by the chemical composition of the liquid; out of the large number of solutions of metallic salts which exist, only a very small proportion yield under any known circumstances thick deposits of reguline metal. It is also largely affected by density of current at the cathode; if the density is very feeble the metal is usually more or less crystalline, if it is moderate the metal is reguline, and much like that which is obtained by ordinary smelting processes, and if it is great the metal separates as a soft black powder in an extremely fine state of division. Even silver may be obtained in this state, and as the limits of range of density within which reguline metal can be obtained are often very narrow, the density of the current at the cathode requires to be very carefully regulated; the range appears to be widest with certain solutions of antimony, copper, and silver. The structure of the metal is also affected by the diffusive power of the liquid and by temperature; some solutions will only yield reguline metal whilst hot.

The layer of metal first deposited is often very different from the succeeding ones obtained by uninterrupted continuance of the current. It often happens, especially with those

FIG. 32.—Nodules of Electro-Deposited Copper.

metals which are difficult to obtain in a reguline state by electrolysis, that at the commencement of deposition the separated metal appears bright and like the ordinary substance; but by continuance of the action it rapidly becomes dull and non-coherent, and passes into the state of dark-coloured or black powder; this happens even with some solutions of copper and silver. If the surface upon which it is deposited is rough, the deposit appears to more readily pass into a state of dark powder. Electro-deposited copper has nearly always a more open texture in its outer layer than in that first deposited, and if a flat disc

FIG. 33.—Silver Crystallised from Weak Solution.

FIG. 34.—Silver Crystallised from Strong Solution.

FIG. 35.—Silver Crystallised from Very Strong Solution

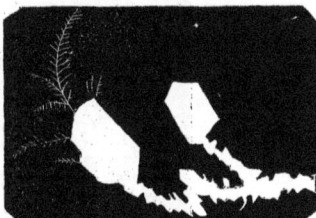

FIG. 36.—Silver Crystallised from Nearly Exhausted Solution.

FIG. 37.—Silver Crystallised from Nearly Exhausted Solution.

of such metal is heated to redness and then cooled, it becomes concave on the side last deposited, owing to the greater contraction in its last deposited layers. Electro-deposited copper articles are very liable to alter greatly in shape when heated to redness.

Every different metal when deposited in what is termed the reguline state has its own peculiarities of structure; thus

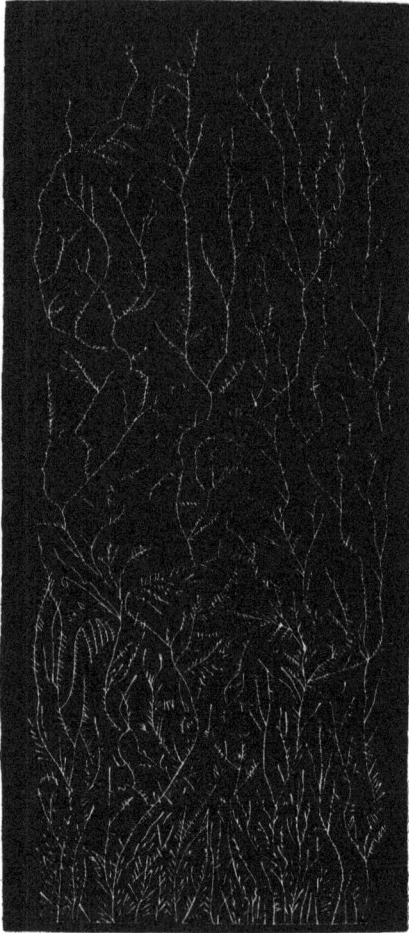

FIG. 38.—Silver Crystallised from 2½ per cent. Solution.
(Magnified four times.)

nickel, when deposited in masses of half an inch or more in thickness, has a surface covered with smooth round knobs, whilst copper, when deposited in thick plates from its sulphate, has a surface covered with rough knobs, with aggredations of nodular masses projecting from the edges sometimes to a distance of several inches. Fig. 32 shows nodular deposits of

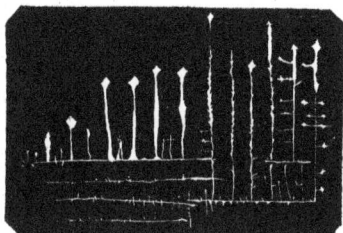

FIG. 39.—Tin Crystallised from a Solution of Stannous Chloride.

FIG. 40.—Gold Crystallised from a Solution of Auric Chloride.

copper from the corners of cathodes in a depositing vat, the largest nodule being 2 inches in length. The greatest length of nodule is in the direction of the greatest density of current and of amount of copper in solution. The preceding figures represent the forms of electro-deposited crystals of silver, tin, and gold. Fig. 33 shows silver as deposited from a weak solution of its nitrate. Fig. 34, ditto, from a strong solution. Fig. 35, from a very strong solution. Figs. 36 and 37, from a nearly exhausted one, and Fig. 38 from a 2½ per cent. solution.

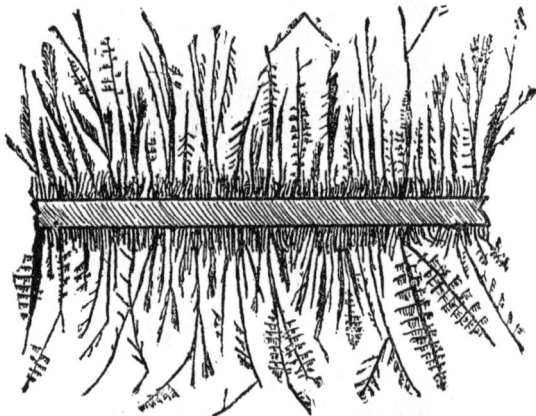

FIG. 41.—Deposit of Lead upon a Strip of Zinc.

FIG. 42.—Deposit of Copper upon a Zinc Wire.

Fig. 39 represents tin from a solution of its chloride, and Fig. 40 gold from a solution of auric chloride (Dr. Gladstone's Lectures, *Telegraphic Journal*, Vol. III., pp. 29—38). Figs. 41 and 42 represent arborisations of lead and copper obtained by electrolytic action (M. C. Decharme, *La Lumière Électrique*, and *Scientific American Supplement*, No. 618, Nov. 5, 1887).

Some electro-copper, deposited under pressure, had a tenacity

of " 23¾ tons per square inch " (*The Electrician*, Vol. XXI.,
p. 431). I have observed that when some thick sheets of copper
were rolled out thin, blisters of all sizes up to two inches in
length, due to air bubbles, appeared all over their surfaces;
these air bubbles may, however, be avoided. According to
Soret, electrolytic copper always contains some hydrogen and
carbonic acid. (*Nature*, Vol. XXXIX., p. 72, and July 4, 1889.)

Necessary Rate of Deposition.—Every different electrolyte,
at every different temperature, must be electrolysed at a
particular rate, within certain limits, in order to continuously
obtain from it the desired quality of metal. The ordinary
copper depositing solution, which consists of a four-fifths saturated and freely acidulated solution of cupric sulphate (*see*
p. 194), yields reguline tough metal when the rate of deposition
does not exceed an increase of thickness of one-twelfth of an
inch (about 2 millimetres) per week of 156 hours ; or three-
eighths of an ounce of copper per square foot per hour, and
when the density of current does not exceed 10 amperes per
square foot of cathode surface. Ten amperes separate 9·84
ounces of copper in 24 hours. (One cubic inch of copper of
specific gravity of 8·95 weighs 2,260 grains, or 5·166 ounces
One avoirdupois ounce equals 437·5 grains. One metre equals
1,000 millimetres, or 39·37 inches. *See* Appendix.)

The following densities of current and electromotive forces
have been given as suitable for depositing different metals from
their usual solutions in a firm, coherent state :—

	Amperes.		Electromo-tive forces.
	Per square decimetre.	Per square foot.	Volts per vat.
Sulphate coppering solution	1·2 to 1·7	11·13 to 15·8	·7
Cyanide ,, ,,	·6	5·56	3·0
,, silvering ,,	·5	4·64	1·0
,, gilding ,,	·1	·928	4·0
Brassing solution	1·0	9·276	3·5
Nickelling ,, ...	·3 to ·6	2·78 to 5·56	2·0 to 3·0
Depositing nickel upon zinc	·4	·372	4·0

("Betrieb der Galvanoplastik mit dynamo-elektrischen
Maschinen zu Zwecken der Graphischen Künste." By Ottomar
Volkmar, 1888, pp. 38-70.)

Electrolytic Separation of Elementary Substances.—These are set free in aqueous solutions. 1st. At the cathode only when they are not readily corroded by the electrolyte, and do not decompose water. Thus all the noble metals, and many of the base ones, are readily separated as cations, but the alkali metals, earth metals, and such metals as manganese, chromium, uranium, tungsten, and molybdenum, are not readily disunited from such liquids. From fused electrolytes, in which water is not present, the metals which decompose water are more easily obtained. And, 2nd, at the anode, in aqueous solutions, some of the elementary gases, such as oxygen and chlorine, which do not readily decompose water, are easily separated ; but fluorine, which instantly acts upon water and liberates its oxygen, cannot be so obtained. Some of the metalloids, such as bromine, iodine, and sulphur, readily separate at the anode in aqueous solutions from some of their compounds, and may thus be obtained in the elementary state. Neither phosphorus nor carbon has been set free by electrolysis in an aqueous solution, but phosphorus has been liberated from fused pyrophosphate of sodium (P. Burckhard, *Chem. News*, Vol. XXI., p. 238), and carbon from a fused mixture of carbonate and silicate of potassium (*Proc.* Birm. Phil. Soc., Vol. IV., 1884, p. 230).

Electrolytic Deposition of Compounds, Mixtures and Alloys.—In many cases the products of electrolysis, especially the secondary ones or those derived from the action of the original ions upon the other bodies present, are compound substances, those at the anode being either acids, oxides, peroxides, salts, or mixtures of gases, and those at the cathode being either alkalies, metallic suboxides, or alloys. When the solution consists of salts of two or more easily and equally reducible metals in suitable proportions, the deposit upon the cathode consists of either a mixture of metals or an alloy. For instance, copper, silver, and gold are easily deposited as reguline alloy from cyanide solutions. Alloys are also sometimes formed by the deposited metal uniting with the metal of the cathode; mercury electro-deposited upon copper forms an alloy.

Secondary Effects of Electrolysis.—The substances which appear at the electrodes are in very many cases not those actually liberated by the current, but products of the action of those substances either upon the electrodes or upon the constituents of the liquid. For instance, in the electrolysis of a solution of a salt of potash with a cathode of platinum the platinum becomes covered with gas, and the liquid at its surface becomes alkaline, because the potassium deposited there decomposes the water, sets free hydrogen, and unites with the oxygen of the water to form potash. In the electrolysis of a solution of nitrate of silver, peroxide of silver appears at the anode as a secondary product, the oxygen liberated there by electrolysis having united with the silver of the salt. Also when iodine is set free at the cathode during electrolysis of a solution of iodic acid, it is a secondary product due to the deoxidising action upon the acid of the hydrogen liberated there by electrolysis; but when set free at the anode during electrolysis of dilute hydriodic acid, it may be viewed either as a direct result of electrolysis or as a secondary one due to the electrolytically liberated oxygen uniting with the hydrogen of the acid and setting free its iodine. In many cases it is difficult to determine with certainty whether a substance which appears at an electrode is due to primary or secondary action. The more impure the solution the greater the risk of secondary effects.

Incidental Phenomena attending Electrolysis.—These are very numerous and varied, almost as much so as the individual substances electrolysed. Not unfrequently gases are evolved at the anode; this occurs most often with insoluble anodes, such as platinum and carbon. Sometimes they are set free at the cathode; this happens when water is electrolysed, and with solutions of highly positive metals, the liberated metal decomposing the water. Frequently the anode dissolves; this occurs when it forms with the acid or oxygen of the liquid a soluble salt. Often it acquires an adherent or non-adherent solid coating; this happens when it is easily corroded and forms an insoluble salt with the oxygen or acid of the liquid; it also occurs when the oxygen set free by electrolysis unites with the metal of the salt in solution to form peroxides, most frequently with salts of lead and silver. In many liquids the anode whilst dissolving partly crumbles and disintegrates to a greater or less extent without dissolving, and the metal falls as a powder or sometimes in large friable pieces to the bottom; with pure silver in dilute hydrofluoric acid this occurs conspicuously, also with antimony in hydrochloric acid, and to a small extent with copper in the ordinary sulphate solution, but least with silver in cyanide of silver plating liquid. Usually the precipitated powder dissolves slowly by chemical action of the liquid. The physical state of the anode affects the amount of resistance and the rapidity of solution; a porous anode dissolves faster than a dense one, largely because it exposes a greater amount of surface to the liquid.

In some instances elementary substances are set free at the anode and dissolve in the liquid; this occurs with iodine liberated from a solution of potassic iodide at an anode of carbon or platinum, similarly with bromine from a solution of potassic bromide. Sometimes the elementary substance liberated unites at once with the water to form a hydrate, which either adheres to the anode or dissolves; this occurs with chlorine set free at a platinum anode in cold hydrochloric acid. In other cases the solid substances formed or set free at the anode are insoluble in the liquid, and either form a coating upon the anode, and impede the current, or fall to the bottom, and are then usually dissolved slowly by the liquid; this happens in an ordinary sulphate of copper solution, in different degrees, when the copper anode contains sulphur, carbon, antimony, arsenic, bismuth, &c.

Frequently a coating of metal is deposited upon the cathode (rarely upon the anode); this occurs with salts of nearly all the easily-reducible metals, especially with gold, silver, platinum, palladium, mercury, copper, nickel, antimony, bismuth, &c., but rarely with solutions of difficultly-reducible ones, such as potassium, sodium, and the alkaline earth metals; with solutions of salts of manganese, uranium, chromium, cobalt, and other easily-oxidisable metals, the deposits upon the cathode sometimes appear as suboxides. When mercury (or a liquid alloy) is used as a cathode, the deposited metal is usually absorbed by it, and largely protected from redissolving in the electrolyte. Even solid cathodes in some cases absorb to a small extent solid metals deposited upon them; for instance, a thin film of zinc deposited upon a perfectly clean and freshly deposited surface of copper, in a cyanide of zinc solution, is absorbed, and imparts to the copper a yellowish appearance ; films of metal also deposited upon other metals, in certain cases, disappear after a time by absorption.

Different metals whilst depositing exhibit different phenomena ; for instance, copper, whilst depositing upon the glass bulb of a thermometer, contracts, compresses the bulb with a force of more than one hundred pounds per square inch, and causes the mercury to rise; cadmium causes it to expand slightly under these circumstances. This phenomenon has been termed "electro-striction." I have observed similar effects by depositing metals upon very fine wires in a stretched condition. Pure grey metallic antimony, whilst being deposited from a solution of tartar-emetic in dilute hydrochloric acid often cracks spontaneously, and curls up into fantastic shapes by its contractile tendency. In these and other cases sounds are evolved; due either to the sudden crackings or to the escape of minute bubbles of hydrogen from the deposited metal ; the latter occurs conspicuously with antimony deposited from its terbromide in an acid solution. With electrodes of mercury in a solution of the double cyanide of that metal and potassium I observed that the surface of the electrode becomes covered with beautiful symmetrical series of waves and a musical sound is emitted. These waves are probably caused by the alternate formation and destruction, by contraction and expansion, of non-conducting films upon the mercury, and are attended by partial intermittance of the current (*Proc.* Roy. Soc., Vol. XII., p. 217).

Many deposited metals contain hydrogen; antimony obtained from an acid solution of its terbromide is often filled with bubbles of it. Deposited hydrogen is very apt to alloy with certain metals which are employed as cathodes; for instance, during the electrolysis of dilute sulphuric acid and certain other liquids, with a cathode of palladium foil, the latter is seen to bend in consequence of absorbing hydrogen, and then when the current is reversed and the palladium made an anode, to unbend by removal of the hydrogen by oxidation. Hydrogen deposited upon steel frequently renders it very brittle, and with every different electrolyte, and every different substance used as an electrode, the phenomena are more or less altered (*see* also pp. 99-105).

Explosive Deposited Antimony.—Not only does the deposited metal sometimes alloy with or penetrate into the mass of the cathode, but in certain special and exceptional cases during the act of deposition it unites with some of the ingredients of the electrolyte and becomes thereby altered in property. For instance, I found that antimony which had been rapidly deposited from a solution of its teroxide in partially diluted hydrochloric acid contained several per cent. of the terchloride derived from the liquid, and possessed a singular explosive property, viz., that if broken or even scratched it suddenly rose in temperature six hundred or more Fahrenheit degrees, according to the massiveness of the piece and the time that had elapsed since it was deposited. When freshly formed this substance has the appearance of highly burnished steel, very different from the colour and appearance of the pure grey metal which has been very slowly deposited from an aqueous solution of tartar-emetic, moderately acidulated with tartaric and hydrochloric acid. But it very gradually changes and loses its singular property (*Phil. Trans.* Roy. Soc., 1857, 1858, 1862).

Purity of Electrolytic Deposits.—It is only in certain cases and in the presence of suitable conditions that a pure metal is deposited from a very impure solution. Substances which are very easily deposited, such as silver, copper and hydrogen, are usually pure, chiefly because it requires a stronger electromotive force or greater density of current to separate most other cations; also through the other substances being usually present in too small a proportion to be at once deposited, or because they are precipitated as insoluble compounds, or through their being themselves insoluble. Even from a very impure solution, if the impurities require very much more energy to separate them, pure metal may be slowly deposited; for instance, copper from a strong solution of mixed salts of the alkali metals.

The separation of a pure metal from an impure solution depends upon several chief conditions :—1. The composition of the electrolyte; 2. The relative amounts of electric energy required to separate the metal, and that required to separate the impurity, and consequently the relative positions of the two metals in thermo-chemical and voltaic series with the particular compound (*see* pp. 39, 50, 56); 3. The strength and density of the current; 4. The degree of concentration of the electrolyte; and, 5. The diffusive power, temperature, and degree of motion of the liquid, &c. As the effect of several of these conditions has been already partly described (*see* pp. 82, 87, 88), the further statements respecting them will be brief.

All electrolytically separated substances, whether anions or cations, are more likely to be pure the greater the degree of purity of the electrolyte; but even from a perfectly pure solution of a metallic salt in pure water, the deposited metal may in certain cases contain hydrogen (*see* p. 105); copper deposited from a perfectly neutral solution of its sulphate is not pure (*see* p. 88). From an impure solution the least electro-positive metal is separated first, and is deposited alone, provided the current is sufficiently weak in proportion to the size of the cathode; for instance, if the ordinary acidified sulphate of copper solution contains any dissolved silver, the silver tends to be deposited first, and by using a large cathode and a weak enough current for a sufficiently long time, and constantly stirring the liquid, the whole of the silver may be deposited first, and in a state of purity; it may also be removed by agitating the liquid with a sufficiently large surface of copper (*see* Arrangement No. 1, p. 74).

The risk of depositing a second metal along with the first one is greater the smaller the proportion of the first metal; the greater that of the second one, the nearer the two metals agree in their amounts of heat of chemical formation in the given compound and of their electromotive force in the particular liquid (*see* pp. 39, 56), and the denser the current is at the cathode. The commonly great degree of purity of electro-deposited copper is chiefly due to the circumstances that it is usually the least electro-positive of all the metals in solution, it is present in very large proportion, and is deposited by a current of not too great a degree of density.

Density of current has great effect. If it is so large that the molecules of the least positive metal at the cathode are deposited faster than fresh ones replace them and those of the impurities are removed, the second least positive metal begins to be deposited along with the first one (*see* also p. 98).

With regard to the influence of diffusion and stirring of the liquid; when an electric current is passed into a cathode in an

electrolyte the immediate layer of liquid in contact with the
cathode is instantly deprived by electro-deposition of every
molecule of the least electro-positive element contained in it,
and unless a fresh supply of the same kind of molecules is in
the interim brought into contact with that surface and those
of the impurities removed, either by the spontaneous diffusive
power of the liquid or by artificial motion, such as stirring, &c.,
molecules of the next least electro-positive element begin to be
separated.   If, therefore, the salt of the second least electro-
positive metal has greater diffusive power than that of the
first one it tends to be more readily decomposed and its metal
deposited.

Decomposability of Electrolytes.—The relative decomposability of electrolytes is important, first, because the less decomposable ones consume more electric energy ; and second, because in electrolysing a mixture of substances the constituents
of the more decomposable ones are liberated first, and this
enables substances to be separated from one another, and impure ones to be purified.   In a mixture the substances first set
free by electrolysis are usually those which require the least
expenditure of energy to liberate them.

The decomposability of a substance is affected by various
circumstances.   Mixtures of liquids usually conduct better and
resist decomposition less than single compounds.   Thus a mixture of oil of vitriol and water transmits a current much more
readily than either the acid or the water alone.   The decomposability of a compound depends largely upon the kind of
liquid in which it is dissolved; for instance, I have observed
that an acidified aqueous solution of teriodide or terbromide of
antimony is easily electrolysed, but that when the same salts are
dissolved in bisulphide of carbon the solutions do not conduct,
and are not decomposed (*see* also p. 87).

Decomposability of the ingredients of a mixture is largely
influenced by the kind of anode ; thus a solution of potassic
cyanide is easily electrolysed with a silver anode, but not with
one of platinum (*see* pp. 91, 94).

The decomposability of an electrolyte appears also to be
affected by the length and resistance of the liquid portion of
the circuit.   This has been shown by Gladstone and Tribe, who

electrolysed water by immersing in it a plate of pure zinc previously coated electrolytically with a loose deposit of spongy copper or platinum, whilst it is known that a plate of zinc and one of copper will not decompose it. They have thus shown that "the dissociation of a binary compound may occur at infinitesimally short distances, when it would not take place when the layer of liquid is enough to offer resistance to the current" (*Proc. Roy. Soc., Vol. XX., p. 219*).

I have observed that a solution of 7 grains of potassic chloride per ounce of water, containing a piece of zinc in contact with a piece of platinum, begins to exhibit a slightly alkaline reaction and deposits oxide of zinc after one or two days, whilst with zinc alone it shows no such effects. The electromotive force of such a combination at 16°C. is about 1·15 volt.

Minimum Electromotive Force Required to Decompose Compounds.—Every electrolyte requires a certain minimum amount of electromotive force in order to decompose it, and the minimum differs widely with different liquids. However great the strength of current no decomposition occurs if the electromotive force is not sufficient. The minimum amount of electric energy usually required to decompose a given compound must be equal to the degree of energy with which its constituents unite, and about equivalent to the quantity of heat evolved by the constituents of that compound when combining together ; if, therefore, we know what that amount of heat is, we can usually calculate the quantity of electric energy required. The amounts of heat evolved in the formation of nearly all ordinary compounds likely to be electrolysed are given in the tables of heat of chemical union on pages 39-42. One volt corresponds to about 22,900 centigrade-gramme calories ; if we know the number of such calories required to decompose a chemical equivalent in grammes of a compound, we have only to divide that number by 22,900 in order to find the minimum number of volts required to electrolyse it ; or if we know the number of volts required to electrolyse it, we have only to multiply that number by 22,900 in order to arrive at the number of such calories. Thus, one gramme of hydrogen, or eight grammes of oxygen, in burning to nine grammes of water, evolve 34,500 calories ; therefore, the number of volts of electromotive force required to electrolyse water is $\dfrac{34500}{22900} = 1\cdot50$ (*see* p. 52).

M. Berthelot, in his experiments on the " Limits of Electrolysis" (*Comptes Rendus*, Vol. XCIII., pp. 661-668), found that it required the electromotive force of at least 4 Daniell cells (= 98,000 centigrade-gramme calories, or 4·28 volts) to deposit potassium into a cathode of mercury from a solution of potassic sulphate with a platinum anode. A solution of iodide of potassium was electrolysed by an electromotive force of 1·16 volt (= 27,000 calories), with liberation of iodine and hydrogen, the acid and base being first separated. One of potassium bromide required 1·746 volt (= 40,000). Chloride of potassium was electrolysed by 2·01 volts (= 46,000 calories), the acid and base being separated, the acid decomposed and gas evolved. Fluoride of potassium required 2·183 volts (= 50,000 calories), oxygen and hydrogen being set free. The haloid salts behaved like the sulphate, being separated into acid and base, and in both cases the minimum electromotive force necessary to produce decomposition was very much less than that required to set free the alkali metal (*Jour.* Chem. Soc., 1882, Vol. XLII., pp. 260, 353). With a solution of ferrous sulphate, the smallest electromotive force capable of effecting electrolysis caused a deposit of iron on the cathode, but there was no evolution of gas at the anode with two Daniell cells or less, the oxygen being used to convert the ferrous into ferric sulphate ; if the electromotive force was greater, oxygen and hydrogen were then set free, with formation of oxide and acid. With manganous sulphate the minimum electromotive force caused a precipitate of manganous dioxide at the anode, hydrogen being evolved at the cathode ; if the electromotive force was then gradually increased, there came a moment at which oxygen was set free at the anode and manganese at the cathode. In all these cases the thermal equivalents of the electromotive forces agreed with those of the chemical actions (*Comptes Rendus*, Vol. XCIII., pp. 757-762 ; *Jour.* Chem. Soc., 1882, Vol. XLII., pp. 260, 353).

Order of Decomposability of Electrolytes.—I have observed that in an aqueous acid mixture hydrochloric acid is electrolysed more readily than water, and water more readily than hydrofluoric acid, also that a solution of selenic acid is electrolysed before one of selenate of nickel. The order of decomposability of compounds is usually the reverse of that of the increase of the amounts of heat they evolve during their formation, as given on pages 39-42. The order of decomposability by electrolysis of aqueous solutions of several of the salts of copper is as follows, the first being the most easily decomposed :—Iodide, bromide, chloride, nitrate, sulphate ; and is indicated by the numbers of centigrade-gramme degrees of heat in the following table :—

Heat of Formation and Solution of Salts of Copper.

	Of Formation.	Of Solution.	Total.
(Cu^2, I^2)	32,520		
(Cu, Br^2)	32,580	$8,250 =$	40,830
(Cu^2, Br^2)	49,970		
(Cu, Cl^2)	51,630	$11,080 =$	62,710
$(Cu, Cl^2, 2H^2O)$	58,500	$4,210 =$	62,710
(Cu^2, Cl^2)	65,750		
$(Cu, O^2, N^2O^4, 6H^2O)$	96,950	$-10,710 =$	86,240
(Cu, O^2, SO^2)	111,490	$15,800 =$	127,290
(Cu, O^2, SO^2, H^2O)	117,950	$9,340 =$	127,290
$(Cu, O^2, SO^2, 5H^2O)$	130,040	$-2,750 =$	127,290

Order of Solution of Anodes and Separation of Ions.—From what has already been said respecting the relations existing between thermal, chemical, voltaic, and electrolytic action, it is manifest that the order of dissolving of a mixture of metals or an alloy as an anode is just the reverse of that of their deposition from a mixed solution of their salts, and that that metal dissolves the first which evolves the most energy, and that anion and cation is separated the first which absorbs the least energy. The order also in which the metals of an alloy dissolve without the aid of an external current is the same as that with it, and the order of deposition of metals from a mixed solution by the simple immersion process is the same as that with an external current. In every case those effects are first produced to which there exists the least amount of resistance.

I 2

Divided Electrolysis.—When an impure liquid or a mixture of liquids is electrolysed either a single substance or several may simultaneously appear at the anode or at the cathode. With a feeble current and large electrodes one substance alone may be set free at each electrode, but by increasing the strength of the current or diminishing the size of one of the electrodes a second or even a third substance may be there liberated, the current dividing its action amongst the various compounds present. Either the least electro-positive or the most abundant cation is liberated at the cathode first, and the next more positive or most abundant ones subsequently as the current density is increased. If the two or more cations are equally electro-positive they are probably each deposited simultaneously in about the same proportions as they exist in solution; but if one of them is a little less positive than the others, that one is the most freely separated; but if it is present in very much smaller proportion it is the least freely separated. By employing, therefore, proportions of the substances larger as their electro-positive property is greater, and sufficient density of current, several or many substances may be simultaneously liberated, notwithstanding that their degrees of strength of chemical affinity are different. For instance, it was by passing a current of great density through moistened hydrate of potash that potassium was first liberated, whilst the much less positive substance, hydrogen, was freely present. To produce great density of current, however, requires considerable electro-motive force.

Of the electrolysis of mixed solutions our knowledge is very imperfect; if the metals dissolved are about equally positive to each other in the particular liquid, and their salts in solution possess nearly equal conducting power, offer equal resistance to reduction, and are present in approximately equal quantities, the current appears to divide itself equally between them, and all the metals are simultaneously deposited. For instance, in a mixed aqueous solution of suitable proportions of the cyanides of zinc and copper, and carbonate of ammonium of a suitable degree of concentration, and at a proper temperature, the electromotive forces of zinc and copper are equal; consequently both the copper and zinc of a brass anode dissolve and deposit equally on the passage of a suitable current.

By electrolysing a mixture of solutions of the sulphates of zinc, cadmium and copper, Favre deposited either one, two, or all three of the metals simultaneously, and concluded that the results vary—1st, with the energy of the current; 2nd, with the electrolytic resistance of the salts; 3rd, with the relative quantity of each salt; and 4th, with the speed of electrolysis (*Comptes Rendus*, Vol. LXXIII.; *Jour.* Chem. Soc., 2nd Series, Vol. X., p. 118). In electrolysing a mixture by means of a current of sufficient electromotive force and density, a portion of the least positive metal is sure to be deposited, even though it be present only in very small amount. The density of current at the anode affects the separation of anions like that at the cathode affects that of cations (*see* also p. 98).

Quantity of Electro-Chemical Action.—The quantities of electrolytic effect at the anode and cathode are always equivalent and differ only in kind. The amount of electrolysis, of substance decomposed, of metal deposited, of anion separated, &c., is directly proportional to the strength of the current or the quantity of electricity passing through the liquid in a given period of time. The amount of electrolysis is also, like the strength of the current, equal at all points in the circuit. This is known as Faraday's Law of Definite Electro-Chemical Action, and may be illustrated as follows:—If four electrolysis vessels are connected in single series, No. 1 containing platinum electrodes, and water acidified by sulphuric acid; No. 2, cyanide of silver plating solution and silver electrodes; No. 3, an acidified solution of blue vitriol and copper electrodes; and No. 4, solution of tartrate of potash and antimony in dilute hydrochloric acid, with antimony electrodes, and a current of electricity of suitable strength be sent through all of them during the same period of time, the amounts of chemical effect, both at the anode and at the cathode, in each of them, will be strictly in proportion to the chemical *equivalents* (not necessarily the atomic weights, *see* p. 46) of the several substances. Thus, in No. 1, for each chemical equivalent, or 1 atomic weight, or 1 part by weight of the monad hydrogen set free at the cathode, an equivalent, or half an atomic weight, or 8 parts by weight of dyad oxygen will be set free at the anode, and an equivalent, or half a molecular weight, or nine parts by weight of water, will be decomposed.   In No. 2, one atomic weight, or 108 parts by weight of monad silver will be deposited at the cathode, and 108 parts be dissolved at the anode, and 1 molecular weight of silver salt be decomposed. In No. 3 half an atomic weight, or 31·5 parts by weight of dyad copper, will be dissolved at the anode, a similar quantity be deposited at the cathode, and half a molecular weight of the copper salt be decomposed.   And in No. 4, antimony in its terchloride being a triad, one-third of an atomic weight, or 40·66 parts by weight of antimony will be deposited at the cathode, a similar weight will unite with 1 atomic weight or 35·5 parts by weight of monad chlorine at the anode, and be dissolved, and one-third of a molecular weight, or 75·5 parts by weight of the terchloride will be decomposed.   By passing an electric current through two liquids in series, one of which yielded pure copper only and the other pure antimony, I found that for each 31·7 parts of copper separated, there were 40·6 parts of antimony deposited, and this agreed with one-third of an equivalent of antimony, being the chemical equivalent of half an atomic weight of dyad copper, or one atomic weight of monad hydrogen, &c.   E. Becquerel obtained a similar result.   In cases where the current simultaneously decomposes two or more salts and liberates two or more substances at an electrode, the equivalent is composed of the collective weights of the several bodies.

E. Becquerel found by electrolysing acidulated water and solutions of cupreous, ferrous and ferric chloride by the same current, that the equivalent amounts of anion determined to the anode were the same in each case, but that the proportions of cation were represented by the following formulæ :—

$$H^2O = Cu^2Cl^2 = FeCl^2 = \frac{Fe^2Cl^6}{3} ;$$ the proportions of iron separated in its proto and persalts, therefore, were as 3 to 2, and the amount of anion it was combined with influenced the amount of iron liberated.   The explanation of the deficiency of deposited metal in solution of the iron persalt is that a portion of the current is expended in separating hydrogen, which reduces some of the salt to the proto state.   The equivalent at the cathode, therefore, in this case is made up of iron and hydrogen.

The electrolysis of persalts is more complex than that of protosalts.   With an incorrodible anode and a copper or iron cathode in an aqueous solution of ferric chloride, chlorine is evolved as gas at the anode, some being dissolved; and at the cathode the effects vary with the density of the current and the strength of the solution. With a feeble current, especially in a strong solution, hydrogen alone is deposited at the cathode, but does not appear in a strong solution, because it is at once consumed in uniting with the chlorine of the persalt and reducing the latter to the proto state. With a strong current, especially in a weak solution, iron is deposited along with the hydrogen, and some of the latter escapes in bubbles. In any case the hydrogen is deposited before the iron, and the stronger the current the greater the proportion of iron deposited to that of the hydrogen.

It is quite possible to obtain a deposit of metallic iron, indirectly and apparently, but not really, from a solution of a persalt of iron, by employing a current of sufficient density at the cathode. In such a liquid, if the current per given area of cathode is small, it is entirely used in doing the easiest work, viz., reducing the persalt to protosalt, and the latter diffuses away as fast as produced; but if it is large, the excess reduces some of the protosalt to metal before it has time to diffuse. Thermo-chemical data (see pp. 39-42) support these conclusions; thus, it takes 13,990 calories to reduce persalt to protosalt, 82,050 to reduce protosalt to metal, and 96,040 to reduce persalt, first to protosalt and then to metal.

By electrolysis, the same quantity of electricity in the form of a current, which will set free 1 atomic weight in grammes (or other denomination of weight) of a monad element will liberate only $\frac{1}{2}$ an atomic weight of a dyad element, $\frac{1}{3}$ of a triad, $\frac{1}{4}$ of a tetrad, &c. The same quantity of current which liberates an equivalent, or 1 part by weight of hydrogen, sets free an equivalent, or 8 parts of oxygen, and decomposes an equivalent, or 9 parts of water, and liberates or decomposes an equivalent quantity of any other suitable substance. And as the chemical equivalent of hydrogen is 1, it is usually employed as the unit in calculating the amount of any other substance liberated or decomposed by electrolysis. Whatever amount of current a particular weight of any metal requires to separate it by electrolysis, that same amount of current will it yield by voltaic action; thus, $\frac{1}{2}$ an atomic weight of dyad zinc yields as

many coulombs of voltaic current as 1 atomic weight of sodium
or other monad metal (*see* p. 64). And conversely, whatever
amount of current a given weight of a particular metal will
produce by voltaic action when forming a protosalt, that same
amount of current is required to separate the same weight of
that metal by electrolytic action. Similar laws are true with
regard to metalloids or other anions. The generating and consuming powers of any metal or metalloid in relation to electric
current are therefore equal, and a single chemical equivalent
of any elementary substance has associated with it the same
amount of electricity.

Notwithstanding that these statements and quantities are
so precise, it is difficult to accurately realise them in experiment, and still more so in large operations; and it is only by
attending to a number of minute precautions that they are
very nearly attained. Usually the loss of the anode is greater
and the gain of the cathode is less than theory indicates, and
than that which is actually due to the current. This chiefly
arises from ordinary chemical action of the liquid, which is not
entirely suspended during the passage of the current (*see* p. 65),
but continues in a modified form, influenced by the presence of
electrolytic products at the surfaces of the electrodes.

Coexistence of Chemical and Electro-Chemical Action.—
Electrolytic actions obey the same general law of equivalence
as ordinary chemical ones, and electro-chemical action by means
of a separate current may be viewed as being ordinary chemical
action in the form of a current, taking place in one large circuit instead of in a multitude of excessively small and immea-
surable ones; and, conversely, ordinary chemical action may be
viewed as being electro-chemical, taking place in an almost
infinite number of such minute circuits. In accordance with
this view, voltaic current has been termed "current affinity."

The electrolytic circuits in which electric currents flow may
be of any degree of magnitude, from the most immeasurably
minute ones to those which flow in the largest electro-metallurgical operations. Currents of all degrees of magnitude may
also coexist and circulate simultaneously in the same metal and
liquid. Thus one large one may flow through the vessel and
outer circuit, whilst local currents circulate between the upper

and lower ends of each electrode (*see* p. 75), and whilst innumerable minute ones in the form of ordinary chemical action circulate all over the electrodes.

Electro-chemical action therefore does not wholly exclude ordinary chemical change (*see* p. 65). In an electrolysis vessel, ordinary chemical actions proceed simultaneously and side by side with the electrolytic ones, and the two kinds of action have frequently opposite effects. Thus, whilst the electric current is always separating the liquid into its constituent parts, chemical action is continually tending to reunite those constituents. For instance, in an acidified solution of cupric sulphate, whilst the current deposits copper upon the cathode chemical action slowly dissolves it off again. In a research on the "Electrolysis of Sulphate of Copper" (*Proc. Birm. Phil. Soc.*, 1882, Vol. III., Part I., pp. 24-80), in which electrodes of sheet copper were employed, I found considerable effects of this kind, especially in hot acidulated solutions (*see* Nos. 2, 4 and 14 of the table below), even with electrodes of pure sheet copper in a solution of very pure cupric sulphate. Comparison plates in similar liquids without an electric current were employed in all cases. The following is a brief summary of a few of the results:—

Differences of Loss of Anode and Gain of Cathode.

No. of Expt.	General Conditions of Experiment.		Loss of Anode. In grains.	Gain of Cathode. In grains	Relative Difference.
1	Pure acidulated solut'n of copper sulphate	Cold	15·35	14·76	1 to ·961
2	Ditto	Hot	17·112	11·16	1 ,, ·652
3	Ditto	Cold	11·556	11·132	1 ,, ·965
4	Ditto	Hot	12·386	9·372	1 ,, ·753
5	Non-acid solution of the pure salt	Cold	4·47	5 685	1 ,, 1·271
6	Ditto	Hot	3·45	4 85	1 ,, 1·406
7	Ordinary acid'lat'd solution	Cold	13 46	12·18	1 ,, ·904
8	Same liquid + 70 gr. of persalt of iron	Cold	14·57	10·72	1 ,, ·735
9	Same liquid	Cold	19·43	16·30	1 ,, ·839
10	Same liquid + 25 gr. of persalt of iron	Cold	19·30	15·29	1 ,, ·792
11	Pure neutral solution	Cold	3·17	3·04	1 ,, ·958
12	Ditto	Hot	2 61	3·25	1 ,, 1·245
13	Pure acidulated solution	Cold	3·30	3·02	1 ,, ·915
14	Ditto	Hot	4·96	·80	1 ,, ·161

Many of these experiments indicate that the two processes, ordinary chemical and electro-chemical, coexist and operate at the same surfaces of liquid and metal; that ordinary chemical action, both of simple oxidation and of corrosion of both electrodes by free acid, takes place in all cases, and is a phenomenon entirely distinct from and independent of electro-chemical corrosion of the anode and deposition upon the cathode. The two classes of phenomena, however, are coincident, and affect each other in various ways. As also the phenomena which occur in fused electrolytes are essentially like those in aqueous ones, similar conclusions may be drawn respecting them. Nos. 7, 8, 9, and 10 show the influence of persalt of iron in reducing the amount of copper deposited.

Many years ago M. E. Becquerel made some experiments to determine the proportion of loss of anode to gain of cathode in depositing copper with copper electrodes. With 5 per cent. of free acid in the solution the loss of anode was always greater than the gain of cathode, and the difference averaged from 2 to 5 per cent. With very pure neutral sulphate solution the loss of anode was sometimes greater and sometimes less than the gain of cathode; the deposited copper was not satisfactory and probably contained oxide; the anode, also, did not dissolve in a clean manner. These results illustrate the necessity of having free acid in the solution.

In consequence of these two actions being essentially independent of each other, an electric current passing out of a piece of copper into a sulphuric acid solution does not directly increase the rapidity of ordinary chemical corrosion of the metal, nor does a current entering from such a liquid into a copper cathode directly or completely protect that metal from such corrosion.

Effects of Stirring the Liquid.—Some of the above experiments show that stirring the liquid increases the ordinary chemical corrosion of the cathode, and therefore that the technical process of swaying to and fro by mechanical means, cathodes which are receiving a deposit in an electrolyte, tends to corrode them; for although the motion removes less diluted liquid, and thus diminishes corrosion, it increases corrosion in a greater degree by bringing fresh particles of corrosive liquid into contact with the electrodes.

As stirring affects the speed of chemical corrosion, so also any chemical or electro-chemical action which causes liquid streams must affect the speed of such corrosion; such streams are produced by both these causes. At the anode both ordinary and electro-chemical action, by causing the formation of soluble salt, and thus increasing the specific gravity of the liquid, cause a downward flow. At the cathode, whilst ordinary corrosion, by forming a soluble salt, increases the specific gravity and tends to cause a downward flow, electro-chemical change, by depositing copper, decreases the specific gravity and causes an upward one, and unless the electric current is very feeble the latter overpowers the former.

Effects of Temperature.—That temperature also, in a much larger degree, influences the corrosion is proved by numerical results. The higher the temperature the greater was the corrosion of the hot copper comparison plates without an electric current and of the hot electrodes; and for equal rise of temperature the increase of corrosion appeared to be greater at the higher temperatures than at the lower ones, because the corrosion itself produced heat. It is more clearly shown in the following tables (*compare* also pp. 66—69) :—

Losses of Copper by Chemical Corrosion in Hot and Cold
Liquids without an Electric Current.

No. of Expt.	Liquid.	In Cold Liquid. Grains.	In Hot Liquid. Grains.	Relative Difference.
1	An acidulated solution of cupric sulphate	·080	4·578	1　　to 57·225
2	Water acidulated with sulphuric acid	·038	1·43	1　　,, 37·63
3	Ditto	·074	·98	1　　,, 11·89
4	Acidulated solution of cupric acetate	·030	·03	1　　,, 1·
5	Acidulated solution of cupric sulphate	·0826	2·056	1　　,, 24·89
6	Ditto	·048	4·355	1　　,, 90·73
7	Ditto	·030	2·43	1　　,, 81·
8	Neutral ditto	·290	·18	1·611,, 1·0
9	Ditto	·270	·15	1·80 ,, 1·0
10	Acidulated ditto	·490	4·38	1　　,, 8·94
11	Water acidulated with sulphuric acid	·15	·94	1　　,, 6·266
12	Ditto	·10	·64	1　　,, 6·4

The ordinary chemical corrosion, therefore, was least affected by heat in the acetate of copper solution, somewhat more so in one of neutral cupric sulphate, still more in acidulated water, and most in acidulated cupric sulphate. (*Proc.* Birm. Phil. Soc., Vol. III., p. 24).

Amounts of Copper Deposited by Equal Currents in Hot and Cold Liquids.

No. of Expt.	Liquid.	In Cold Liquid. In grains.	In Hot Liquid. In grains.	Relative Difference.
1	Acidified solution of cupric sulphate......	16·107	14·8	1 to ·919
2	Ditto	25·561	21·624	1 ,, ·845
3	Acidified solution of cupric acetate	·7	1·09	1 ,, 1·557
4	Neutral solution of cupric sulphate......	14·76	11·16	1 ,, ·756
5	Ditto	11·132	9·272	1 ,, ·833
6	Ditto	5·685	4·85	1 ,, ·853
7	Ditto	3·04	3·25	1 ,, 1·069
8	Acidified solution of pure cupric sulphate	3·02	·8	1 ,, ·265

Electrolytic Balance of Chemical and Electro-Chemical Action. — As in all electrolytes which by ordinary chemical action corrode and dissolve the deposited metal, the amount of metal deposited is less than equivalent to the quantity of current which has passed, it is evident that in all such liquids, by commencing the electrolysis with a very small current, no deposit will appear until the current attains a certain degree of magnitude, viz., that at which the rate of ordinary chemical corrosion and that of electro-deposition are equal, and the two opposite actions just balance each other.

In an experimental research on "The Electrolytic Balance of Chemical Corrosion" (*Proc.* Birm. Phil. Soc., 1882, Vol. III., pp. 268-304) I have shown that this state of balance is influenced by at least seven or eight different circumstances, viz., ordinary chemical corrosion, strength of current, nature of cathode, temperature, proportion of water, dissolved metallic salt, free acid, and of any soluble impurities present, and that probably all these have numerical values. In one experiment

a rise of temperature of the liquid from 60° to 120°F., or of 60 F.
degrees, was balanced by an increased strength of current from
·002306 to ·003282 ampere, or of ·000976 ampere, equal to a
difference of 37·6 per cent. The cathode and electrolyte employed in these experiments were silver, and the electrolyte
was a solution of argentic cyanide containing a small amount of
free potassic cyanide; but the method of experiment is applicable to other metals and electrolytes. Many other cases in
which electro-chemical is balanced by ordinary chemical action
remain to be examined.

Any mixed electrolyte with a current passing through it and
setting free only one of its cations in the form of a corrodible
metal, and any of the conditions being then so altered that a
second cation begins to be deposited, constitutes an example of
an "electrolytic balance." Such a case is that of the ordinary
acidified solution of cupric sulphate used for refining copper,
with the current increased or some other condition so altered
that a second metal begins also to be deposited.

The same subject has also been further examined by me in
a research on "The Chemical Corrosion of Cathodes," and the
results published (*Proc.* Birm. Phil. Soc., 1882, Vol. III.,
pp. 305-324; *The Electrician*, Vol. XI., p. 213). In some cases
the rate of corrosion of a cathode was found to be increased
during electrolysis, in consequence of evolution of hydrogen
and the consequent upward motion of the liquid, bringing
fresh corrosive particles into contact with the cathode.

As in an electrolysis vessel, with a warm corrosive solution
of a metallic salt, a considerable amount of current may pass
before metal commences to be deposited, so, conversely, in a
heated voltaic cell, a large proportion of the positive metal
may dissolve by ordinary chemical action without contributing
to the external or useful current (*see* pp. 65, 69).

Electro-Chemical Equivalents of Substances.—From the
results obtained by several experimentalists it is evidently
possible, under very limited conditions, especially by employing silver electrodes in a solution of argentic nitrate, to obtain
very nearly as much metal deposited upon the cathode as is
dissolved from the anode, and very nearly the equivalent of
metal deposited to that of electric current passed. It has

been found, by very carefully made experiments, that a current
of a strength of one ampere, passed through dilute sulphuric
acid during one second, liberates about ·00001035 gramme
weight of hydrogen (= ·0373 gramme per hour) ; or if passed
through a solution of cupric sulphate during one hour, sets free

Electro-Chemical Equivalents of Elementary Substances.

Substance.	Equivalent Weight.	Grammes per Coulomb.	Grammes per Ampere per hour.
Aluminium ...	9·1	·0000942	·33912
Antimony	40·66	·000414	1·4904
Bromine.........	79·75	·0008282	2·9815
Cadmium	55·8	·0005775	2·079
Calcium	19·95	·0002068	·74448
Carbon	3·	·00003105	·11178
Chlorine.........	35·5	·0003675	1·323
Cobalt...	29·3	·0003054	1 0994
Copper	31·5	·0003261	1 17396
Fluorine.........	19·	·0001966	·70776
Gold	65·4	·000678	2·5128
Hydrogen	1·	·00001035	·037296
Iodine............	126·53	·0013147	4·7329
Iron	27·95	·0002898	1·04328
,, in persalts.	18·64	·0001929	·69444
Lead	103·2	·0010684	3·8462
Lithium	7·01	·00007245	·26082
Magnesium ...	11·97	·00011567	·41741
Manganese......	18·27	·0001891	·58076
Mercury.........	99·9	·0010351	3·72636
Nickel............	29·3	·0003054	1·0994
Nitrogen.........	4·3	·0000445	·1602
Oxygen	7·98	·0000828	·29808
Potassium	39·04	·0004047	1·4569
Silicon	14·	·0001449	·52164
Silver	107·66	·001118	4·0248
Sodium	22·99	·0002381	·85716
Sulphur	16·	·0001656	·59616
Tin	58·9	·0006096	2·19456
Zinc..............	32·45	·0003364	1·21104

about 1·18656 gramme of copper; or if through one of argentic
nitrate during that time, deposits about 4·025 grammes (or
62·1138 grains) of silver. (One gramme equals 15·432 grains.)
According to Dr. H. Hammerl, with a solution of cupric sulphate, " the greatest permissible strength of current, for
which the deposit may be safely assumed to be a measure of

the current, is about ·7 ampere per square decimetre of the cathode surface" (*Nature*, Vol. XXIX., p. 227). This equals 63 amperes per square foot, and is probably too large a number (*see* pp. 207, 208).

The quantity by weight of an ion separated at an electrode in a given period of time is proportional to the strength of the current, and may be ascertained by multiplying the chemical equivalent weight of the ion by the weight of hydrogen separated in the same time by the same amount of current. The quantities thus arrived at are called " electro-chemical or electrolytic equivalents."

The table on page 126 contains the "electro-chemical equivalents," or quantities of various elementary substances separated by one coulomb of current, or by one ampere strength of current in one second ; many of these numbers have been verified by experiment. To deposit one gramme of copper requires 3,066 coulombs.

Secondary Products of Electrolysis.—In many cases, instead of elementary substances being set free at both electrodes, the passage of the current is attended by the formation of equivalent amounts of chemical compounds at the anode, and the decomposition of equivalent quantities of such compounds at the cathode ; this occurs in a solution of cupric sulphate with a copper anode, and in many other instances. It is usually when a non-corroded substance is used as an anode that an elementary body is set free at that electrode. Sometimes, however, the elementary substance liberated acts chemically upon the water or other bodies present and forms new compounds. Various instances of this kind have already been mentioned.

Electro-Chemical Equivalents of Compound Substances.— The passage of a definite quantity of electric current through an electrolyte, not only liberates strictly proportional quantities of elementary substances at the anode and cathode, but also decomposes a definite and equivalent amount of the compound in the electrolyte. The following table contains the chemical formulæ, molecular and equivalent weights of a number of such compounds, together with the quantities of them decomposed by one coulomb of current. To find the amount of each compound decomposed by one coulomb, multiply the chemical equivalent of the compound by the weight in grammes of hydrogen separated by the same quantity of current :—

Electro-Chemical Equivalents of Compound Substances.

Substance.	Formulæ.	Molecular Weight.	Equivalent Weight.	Grammes decomposed per coulomb.	Gr'mmes per ampere per hour.
Water	H^2O	17·96	8·98	·00009315	·3348
Hydrofluoric acid	HF	20·0	20·0	·00020695	·7452
Hydrochloric ,, ...	HCl	36·5	36·5	·00037785	1·3601
Carbonic ,, ...	CO^2	43·89	21·95	·00022768	·8·965
Nitric ,, ...	HNO^3	62·88	62·88	·00065206	2·3472
Hydrosulphuric acid...	H^2S	33·98	16·99	·00017618	·6343
Sulphurous ,, ...	H^2SO^3	81·86	40·93	·000·2444	1·5280
Sulphuric ,, ...	H^2SO^4	97·82	48·91	·00050719	1·8259
Phosphoric ,, ...	H^3PO^4	97·80	32·9	·00034117	1·2312
Arsenious ,, ...	As^2O^3	197·68	65·89	·00068327	2·4599
Arsenic ,, ...	As^2O^5	229·6	76·5	·0007933	2·8548
Terchloride of antimony	$SbCl^3$	228·5	76·2	·00079019	2·8447
,, bismuth.	$BiCl^3$	316·5	105·5	·00112403	4·0984
Nitrate of ,,	$Bi3NO^3+5H^2O$	485·44	161·81	·00167473	6·0290
Chloride of platinum...	$PtCl^4$	339·0	84·75	·00087857	3·1630
Terchloride of gold.....	$AuCl^3$	303·0	101·0	·00104737	3·7706
Nitrate of silver......	$AgNO^3$	169·54	169·54	·00175813	6·3288
Fluoride ,, ...	AgF	126·66	126·66	·00131346	4·7268
Chloride ,, ...	$AgCl$	143·16	143·16	·00148457	5·3446
Iodide ,, ...	AgI	234·19	234·19	·00242855	8·7430
Mercuric chloride	$HgCl^2$	270·8	135·4	·00140409	5·0548
Cupric nitrate	$Cu2NO^3+6H^2O$	294·52	147·26	·00152708	5·4975
Cupreous chloride......	Cu^2Cl^2	198·	99·	·00102693	3·6970
Cupric ,, ...	$CuCl^2$	133·74	66·87	·00069344	2·4962
,, sulphate	$CuSO^4+5H^2O$	249·5	124·75	·00129366	4·6573
Chloride of nickel......	$NiCl^2$	130·	65·	·00067405	2·4264
Sulphate ,, ...	$NiSO^4+7H^2O$	281·	140·5	·00145698	5·2452
Nitrate of cobalt......	$Co2NO^3+6H^2O$	291·	145·5	·00150583	5·4324
Chloride ,, ...	$CoCl^2$	130·	65·	·00067405	2·4264
Sulphate ,, ...	$CoSO^4+7H^2O$	281·	140·5	·00145698	5·2452
Ferrous chloride	$FeCl^2$	127·	63·5	·00065849	2·3706
Ferric ,, ...	Fe^2Cl^6	325·	54·1	·00055994	2·0158
Ferrous sulphate	$FeSO^4+7H^2O$	278·	139·	·00144143	5·1890
Manganous chloride...	$MnCl^2+4H^2O$	198·	99·	·00102663	3·6956
,, sulphate..	$MnSO^4+5H^2O$	241·	120·5	·00124958	4·4986
Nitrate of lead	$Pb2NO^3$	331·	165·5	·00171623	6·1776
Sulphate of thallium .	Tl^2SO^4	504·	252·0	·00261324	9·4068
Stannous chloride......	$SnCl^2$	189·0	94·5	·00097996	3·5280
Chloride of cadmium	$CdCl^2+2H^2O$	219·0	109·5	·00113551	4·0878
,, zinc	$ZnCl^2$	136·0	68·0	·00070516	2·5380
Sulphate of zinc	$ZnSO^4+7H^2O$	287·0	143·5	·00148809	5·3568
Chloride of magnesium	$MgCl^2$	95·0	47·5	·00049257	1·7734
Sulphate ,,	$MgSO^4+7H^2O$	226·0	113·0	·00117181	4·2185
Cryolite	$6NaF, Al^2F^6$	421·0	70·16	·00072756	2·6194
Chloride of aluminium	Al^2Cl^6	268·0	44·7	·00046354	1·6686
Sodio-chloride ,,	$2NaCl, Al^2Cl^6$	382·0	63·7	·00066057	2·3782
Potash alum	$KAl2SO^4+12H^2O$	473·7	118·42	·00122801	4·4208
Chloride of calcium ...	$CaCl^2$	110·0	55·0	·00057035	2·0533

Electro-Chemical Equivalents of Compound Substances
(continued).

Substance.	Formulæ.	Molecular Weight.	Equivalent Weight.	Grammes decomposed per coulomb.	Gr'mmes per ampere per hour.
Caustic lime	CaO	56·0	28·0	·00029036	1·0440
,, soda	NaHO	39·86	19·93	·00020667	·7440
Chloride of sodium ...	NaCl	58·4	58·4	·00060561	2·1802
Sulphate ,, ..	Na^{2}SO^4+10H^{2}O	321·44	160·72	·00166666	5·9976
Carbonate ,, ...	Na^{2}CO^3+10H^{2}O	285·48	142·74	·00148021	5·32c0
Phosphate ,, ...	Na^{2}HPO^4+12H^{2}O	293·36	97·79	·001014C8	3·6508
Caustic potash	KHO	56·1	28·05	·00029088	1·0476
Nitrate of potassium..	KNO^3	101·0	101·0	·00104737	3·7686
Chloride ,,	KCl	74·6	74·6	·00077360	2·7850
Chlorate ,, ..	KClO^3	122·5	122·5	·00127032	4·5720
Bromide ,, ..	KBr	119·1	119·1	·00123506	4·4460
Iodide ,, ..	KI	166·0	166·0	·00172143	6·1970
Sulphate ,, ..	K^{2}SO^4	174·	87·	·00090219	3·3192
Carbonate ,, ..	K^{2}CO^3	138·	69·	·00071553	2·5758
Ammonia..............	H^{3}N	17·0	5·66	·00005869	·2113
Nitrate of ammonium	H^{4}N, NO^3	80·0	80·0	·00082960	2·9867
Chloride ,,	H^{4}N, Cl	53·5	53·5	·00055479	1·9973
Sulphate ,,	(2H^{4}N), SO^4	132·0	66·	·00068444	2·4638
Hydrocyanic acid	HC^{2}N	27·	27·	·00027999	1·0008
Cyanide of potassium.	KC^{2}N	65·0	65·0	·00068405	2·4624

Some of the above equivalent weights are inferential, and have not been tested by actual electrolysis.

Consumption of Electric Energy in Electrolysis.—Not only precise quantities of current but definite amounts of electric energy are required to decompose or separate given weights of substances. The amount of electric energy required to separate from a compound a given weight of metal or metalloid depends essentially, 1st, upon the strength of chemical union to be overcome; 2nd, the chemical equivalent of the substance to be separated; and 3rd, the amount of conduction resistance of the electrolyte. It may also depend incidentally and largely upon other circumstances. For instance, whether the electrolyte has to be maintained in the liquid state by heat of conduction-resistance of the current, as in the case of separation of aluminium from cryolite; in such case a further large amount of energy is consumed in melting the substance, and in compensating the continual loss of heat caused by cooling influences. According to M. Grotian, the

K

overcoming of the friction of the molecules is an essential part
of the work done by an electrolysing current (*Nature*, Vol. XIV.,
p. 142).

It is usually considered that in every case of electrolysis,
where the same compound is formed at the anode as is decomposed at the cathode, there is no counter electromotive force of
polarisation to be overcome in the electrolyte, the algebraic sum
of all the electromotive forces in the chemical changes which
occur is nil, and the amount of energy consumed in pure
electrolytic decomposition is zero; also that with copper electrodes in a saturated solution of cupric sulphate, the feeblest
current will deposit copper. This, however, is only true provided the solution is uniform and of proper composition,
and the current is not too strong. We know that in consequence of the corrosive action of the liquid, it is very difficult to arrange the electro-deposition of copper from a neutral
saturated solution of its sulphate with electrodes of pure copper
so that the very feeblest current deposits copper (*see* p. 124).
As the amount of energy evolved at a copper anode in a pure
acidulated solution of cupric sulphate is exactly equal to that
absorbed at the cathode, and as the quantity of copper deposited is very nearly equal to that dissolved, and is in nearly
the same physical state, only a very small amount of electric
energy is expended in the pure act of electrolysis, and the
chief amount consumed is that used in overcoming conduction
resistance, producing heat in the electrolyte, balancing chemical corrosion, and conveying the metal from one electrode
to the other.

In some cases, where the anode is not dissolved, but an elementary anion is liberated, an extra and large amount of energy
disappears, equivalent to the amount of heat evolved by the
chemical union of the elementary substances forming the compound which has been decomposed by the current. This occurs
when the ordinary cupric sulphate solution is electrolysed with
an insoluble anode of platinum, an extra amount of energy is
then absorbed and becomes chemically potential in the oxygen
separated at the anode, and this amount is equivalent to the
quantity of heat evolved by the chemical union of that quantity
of oxygen with its equivalent of metallic copper to form cupric
oxide.

Minimum Amounts of Energy Necessary to Decompose Compounds.—The minimum amount of electric energy required to decompose a given compound must be at least equivalent to the quantity of heat produced by the separated substances when reuniting to form the original body (*see* pp. 39-42) ; and it may be much more according to circumstances. If there is any polarisation or counter electromotive force in the electrolysis vessel, that may require much more energy ; the conduction resistance of the liquid also requires some ; a portion is consumed in transporting the ions ; some is lost by secondary chemical actions ; the losses in this way, especially in mixed or impure solutions, are often difficult to estimate, and the actual consumption of energy in most cases can only be ascertained with certainty by means of experiments.

Theoretical Amounts of Energy required to Decompose Compounds.

Substances Decomposed.		Substances Separated.		Thermal Energy Required.	Mechanical Energy Required.	Number of Horsepower during one minute.	Number of Watts during one minute.
	Molecular Weights.		Grammes.	*a* C.G. Calories.	*b* Foot pounds.	*c*	*d*
(H², O)	18·0	H O	2· 16·	68360·	209588·	6·351	4737·85
(H, Cl)	36·37	H Cl	1· 35·37	39315·	120540·	3·6524	2724·92
(H, Br)	80·75	H Br	1· 79·75	28380·	87013·	2·6367	1967·01
(H, I)	127·5	H I	1· 126·5	13170·	40379·	1·2236	912·81
(Sb, Cl³)	228·1	Sb Cl	122· 106·1	91390·	280213·	8·4913	6334·49
(Bi, Cl³)	316·1	Bi Cl	210· 106·1	90630·	277872·	8·4203	6281·12
(Ag², Cl²)	286·86	Ag Cl	215·3 70·74	58760·	180158·	5·4590	4072·66
(Hg, Cl²)	270·5	Hg Cl	199·8 70·7	63160·	193649·	5·8681	4377·62
(Cu, O)	79·	Cu O	63· 16·	37160·	113933·	3·4525	2575·56
(Cu, Cl²)	133·7	Cu Cl	63· 70·7	62710·	192269·	5·8263	4346·43
(Cu, Br²)	222·5	Cu Br	63· 159·5	40836·	125203·	3·7940	2830·34
(Cu, O², SO²)	149·	Cu O SO²	63· 32· 64·	111490·	341828·	10·3584	7727·38

K 2

Theoretical Amounts of Energy required to Decompose Compounds
(continued).

Substances Decomposed.		Substances Separated.		Thermal Energy Required.	Mechanical Energy Required.	Number of Horse-power during one minute.	Number of Watts during one minute.
	Molecular Weights.		Grammes.	a. C.G. Calories.	b. Foot pounds.	c	d
$(Cu, O^2, SO^2 + Aq)$		Cu O^2 SO^2	63· 32· 64·	127290·	389965·	11·817	8815·48
(Ni, Cl^2)	129·3	Ni Cl	58·6 70·7	93700·	287284·	8·7056	6494·35
(Co, Cl^2)	129·3	Co Cl	58·6 70·7	97820·	299916·	9·0884	6779·91
(Fe, Cl^2)	126·6	Fe Cl	55·9 70·7	99950·	306447·	9·2863	6927·77
(Fe^2, Cl^6)	323·9	Fe Cl	111·8 212·1	255440·	783179·	23·7327	17704·56
(Mn, Cl^2)	125·5	Mn Cl	54·8 70·7	128000·	392448·	11·8924	8871·69
(Mn, O^2, SO^2)	150·8	Mn. O SO^2	54·8 32· 64·	178790·	548170·	16·6112	12391·94
(Al^2, Cl^6)	266·	Al Cl	54·6 212·1	321960·	987129·	29·9130	22315·07
(Pb, Cl^2)	277·1	Pb Cl	206·4 70·7	82770·	253773·	7·6901	5737·02
(Pb, O)	222·4	Pb O	206·4 16·	50300·	154220·	4·6733	3486·29
(Sn, Cl^2)	188·5	Sn Cl	117·8 70·7	80790·	247702·	7·5061	5599·56
(Cd, Cl^2)	182·3	Cd Cl	111·6 70·7	93240·	285874·	8·6628	6462·46
(Cd, O^2, SO^2)	207·6	Cd O SO^2	111·6 32· 64·	150470·	461341·	13·980	10429·08
(Zn, Cl^2)	135·6	Zn Cl	64·9 70·7	97210·	298046·	9·0317	6767·63
$(Zn, O^2, SO^2 + Aq)$		Zn O^2 SO^2	64·9 32· 64·	177420·	543544·	16·471	12287·37
(Sr, Cl^2)	157.9	Sr Cl	87·2 70·7	184550·	565830·	17·1464	12791·17
(Ba, Cl^2)	207·5	Ba Cl	136·8 70·7	197770·	597165·	18·0959	13499·52
(Na^2, Cl^2)	116·7	Na Cl	46· 70·7	195380·	599035·	18·1526	13541·81
(K^2, Cl^2)	148·7	K Cl	78· 70·7	211220·	647601·	19·6242	14639·67
(K^2, Cy^2)	130·	K Cy	78· 52·	130700·	400726·	12·1432	9058·82

The preceding table shows the minimum theoretical amounts of electric energy in watts, and mechanical energy in foot-pounds and minutes of horse-power, required to decompose molecular weights in grammes of various compounds, and separate their constituents by electrolysis, as calculated from the amounts of thermal energy in centigrade-gramme calories (*see* pp. 39-42) evolved by the same weights of the same constituents when uniting together to form the same compounds in the same physical state. Many of the numbers in columns *b*, *c*, and *d* require to be verified by experiment.

Although the numbers in the table are so precise, they probably do not in any case represent the whole, nor in many cases anything like the whole, of the actual amount of electric energy consumed in decomposing the respective substances ; and the amounts of product obtained are often much less than those stated. In some cases the anode is corroded and the anion is not entirely set free; in others the anion unites chemically with the liquid, or it diffuses through the solution and unites either with the liberated cation or with the cathode; and in others the cation is either absorbed by the cathode, or diffuses through the liquid and reunites with the anion, or it is oxidised by the water, and hydrogen evolved in its stead ; or no visible deposit at all occurs, the hydrogen being expended in deoxidising a persalt present. Other sources of loss have already been mentioned, especially that by chemical corrosion of the deposited metal (pp. 65-69, 120-125). Nevertheless, the numbers are valuable as constituting one of the chief data required in estimating the total energy expended in the instances given.

An experiment made by Dr. J. Hopkinson on the electrolytic separation of aluminium from fused cryolite, by means of carbon electrodes, according to Dr. Kleiner's process, illustrates the difference between the theoretical and actual amounts of a metal obtained from its compounds by electrolysis. "The mean current was 100·2 amperes, the mean potential 57·43 volts, and the mean energy was 5604·2 watts. The duration of the experiment was 10,380 seconds, and the total electric force applied was equal to 21·6 horse-power hour. The aluminium produced was 60 grammes, the theoretical equivalent of current passed being 93·09 grammes." The chemical formula of cryolite is

3NaF, AlF³. The 35·55 per cent. excess of energy required in this case was chiefly expended in producing heat.

Total Consumption of Electric Energy in Electrolysis.—In attempting to ascertain the total consumption of electric energy in depositing a given weight of metal we have to consider the following items :—(1), the amount converted into heat by conduction resistance throughout the circuit; (2), the quantity lost in replacing metal lost by corrosion of the cathode—this is often considerable, and may amount in the case of copper to a large percentage if the solution is hot or a persalt of iron is dissolved in the liquid (*see* pp. 65, 120); in depositing zinc it would probably be much greater ; (3), the loss by overcoming polarisation, and any other counter-electromotive force, if there is any, at the surfaces of the electrodes (p. 95) ; (4), the loss incurred in transporting the ions and in causing visible movements in the electrolyte; and (5), incidental losses arising from secondary actions. On the large scale there is, in addition to these, often considerable losses by accidental short circuiting and leakages of current. As we cannot correctly estimate all these, the best way is to determine by actual experiment the total consumption of energy, by measuring the number of volts and amperes expended, and weighing the amount of metal finally deposited.

Loss of Energy as Heat of Conduction-Resistance.—According to R. Sabine, the table on page 135 shows the losses of horse-power per 1,000 yards of conductor, and elevations of temperature which occur on passing an electric current through wires and rods of bright copper.

Theories of Electrolysis.—The chief facts of electrolysis having been now described the reader will be better able to judge respecting the reasonableness of any theory put forward to explain them. Various theories have been proposed, but none have been very clear or satisfactory ; one of the best known is that propounded by Faraday. He considered that electrolysis resulted from a peculiar corpuscular action developed in the direction of the current, and that it proceeded from a force which was either added to the affinity of the bodies present or determined the direction of that force. That the electrolyte

Loss of Electric Energy and Rise of Temperature in Conductors of Bright Copper 1,000 yards long.—(R. SABINE.)

AMPERES.

Diameter of Conductor (Inch)	5 H.P.	5 Fah.	10 H.P.	10 Fah.	20 H.P.	20 Fah.	30 H.P.	30 Fah.	40 H.P.	40 Fah.	50 H.P.	50 Fah.	60 H.P.	60 Fah.	70 H.P.	70 Fah.	80 H.P.	80 Fah.	90 H.P.	90 Fah.	100 H.P.	100 Fah.	200 H.P.	200 Fah.	300 H.P.	300 Fah.	400 H.P.	400 Fah.	500 H.P.	500 Fah.	600 H.P.	600 Fah.	700 H.P.	700 Fah.	800 H.P.	800 Fah.	900 H.P.	900 Fah.	1000 H.P.	1000 Fah.
·04	·89	38	7·1	29																																				
·05	·49	19	·76	18	·24	49																																		
·06	·36	12	·55	12	·19	36	·31	47																																
·07	·24		·45	9	·15	27	·20	36	·51	51																														
·08	·18		·36	7	·13	21	·19	25	·41	41	·30	31																												
·09	·15		·31	5	·11	17	·13	19	·34	18·3	·18	18																												
·10	·11		·26	4				13	·24		·67	5																												
·11	·11		·22	3				5	·43		·45	4	·97	7																										
·12	·09		·19	2			·44	6	·34	2	·53	3	·72	4	·94	5																								
·13	·076		·11	1			·33	1	·29	1	·45	2	·64	3	·88	4	·95	3																						
·14				1			·24	1	·24	1	·37	1	·53	2	·73	2	·70	3	·87	3																				
·15					·20	1	·18				·28	1	·40	1	·54	1	·64	2	·72	2	·78	2																		
·20					·15	1							·36		·49	1	·57		·62	2	·45	1																		
·25															·45	1					47·3	49	9·3	37																
·30																					26·48	29	7·7	29	11·18	40														
·35																					17·34	18	5·8	22	·98	28														
·40																					12·28	13	4·9	16	·73		16·84	41												
·45																					8·21	10	4·2	14			13·33	32												
·50																					6·18	8	3·9	12	9·1			18	21·9	50										
·55																					4·14	6	3·1	9	7·3		8		14·1	21	77·39									
·60																					3·10	4	1·8	3	4·1		7·5	14	11·8	18	10·5	16	14·6	22	19·4	29	25·1	36	31·6	44
·65																					·97	3			2·6		4·3	7	7·3	8	11·0		9·9	14	14·2	18	16·6	23	20·9	2
·70																					·78	2					3·4	5	5·0	7	7·2	10	7·0	10	9·0	13	11·7	16	14·6	2
·75																					·45	1					4·7		3·5	5	5·0	7	5·2	7	7·0	9	8·8	11	10·9	14
1·00																															3·8	5	5·5		5·5	6	7·1	7	8·8	9
1·25																																	5·2		5·2	7				
1·50																																								
1·75																																								
2·00																																								
2·25																																								

consisted of a mass of acting particles, of which all that were in the course of the current contributed to the terminal action, and in consequence of the affinity between the elements being weakened or partially neutralised by the current parallel to its own course in one direction, and strengthened and assisted in the other, the combined particles acquired a tendency to move in different directions. The particles of one element, *a*, cannot travel from one pole to the other unless they meet with particles of an opposed substance, *b*, ready to move in the opposite direction; for in consequence of their increased affinity for those particles, and the diminution of their affinity for those which they have left behind, they are continually driven forward.

Any tolerably complete theory of electrolysis of a fundamental character, must be a mechanical one, based upon the assumption of molecular motion, or something equivalent to it, and be expressible in mathematical and geometrical terms. Whilst also the theory must represent the kind of molecular or etherial motion constituting an electric current, it must also be consistent with all the numerous and varied phenomena attending electrolysis, and as the essential kinds of molecular motion which occur at the two electrodes respectively, are probably more or less modified in the case of every different electrode and electrolyte, a complete theory must admit of extremely varied applications.

Clausius suggests that the atoms or groups of atoms constituting a molecule, revolve around one another similarly to planets, and are sometimes nearer to, and sometimes farther from, each other (Poggendorff's *Annalen*, Vol. CLVI., pp. 618-626). Some scientific men believe that an atom of a single elementary substance is more complex than the whole solar system, and that all its parts are in motion. Favre states that in each voltaic couple the molecules are electrolysed *successively*, and that when the absolute number of vibrations which correspond to a given intensity have been determined, the *absolute weight* of the chemical molecules will be known (*Comptes Rendus Acad. Sci.*, Vol. LXXIII., p. 971; *Jour. Chem. Soc.*, 2nd Series, Vol. X., p. 25).

The immediate or primary electrolytic changes are evidently results of molecular energy transmitted either along the wires or its surrounding dielectric from the source of the current; and the energy so transmitted is substantially the same in its chief properties and electrolytic effects, whether it proceeds from a voltaic battery, a thermo-pile, or a dynamo-electric machine. Any theory which explains electrolysis must also be consistent with the fact that in the act of electrolysis the homogeneous dynamic-electric energy is converted into potential molecular energy, as varied in kind as the properties of the liberated elementary substances. It must also explain why, in certain cases, the same element may be an anion in one combination and a cation in another.

SECTION F.*

THE GENERATION OF ELECTRIC CURRENTS BY DYNAMO MACHINES.

A DYNAMO is a machine for the conversion of mechanical into electrical energy. This is accomplished by the rotation in a certain definite manner of a conducting circuit—or part of such circuit—in a magnetic field. The magnetic field may be due either to permanent steel magnets or to electro-magnets—in modern dynamos the latter method is invariably employed. In order to explain what is meant by a "magnetic field" we may refer to the well-known experiment in which iron filings are sprinkled upon a sheet of paper or glass placed over a magnet, and which are found to arrange themselves in certain definite lines. If the magnet be a straight bar of steel those lines may be roughly described as extending outward from one end and bending round in symmetrical curves to the other end of the magnet; other lines lying within these leave the magnet on either side, and enter at symmetrical points on the same side towards the opposite end. If, however, the magnet be of a horseshoe shape, then by far the greater number of lines pass nearly straight across from side to side between the poles or ends of the magnet. This straight-line type of magnetic field is the kind which is chiefly employed in dynamo-electric machinery, and it is obtained whenever two poles of opposite sign are placed face to face.

The existence of the lines of force which emanate from a magnet is made apparent in the space external to the poles by means of the experiment with iron filings referred to above; but it can also and as easily be demonstrated that these lines form closed curves through the bar itself. If a glass tube be filled with iron filings, and its contents be converted into an

---

* This Section is written by **Mr. Guy C. Fricker.**

electro-magnet by the passage of a current through a wire coiled
helically around the tube, the filings may be seen to arrange
themselves longitudinally within it, so as to close and complete
the external lines indicated by the filings strewn on a card in
the first experiment. In an electro-magnet the number of
lines increases with the exciting current—at first almost in
direct proportion to it, but after the magnetic induction in the
iron has reached a certain point, in a gradually diminishing
ratio—until a state of " saturation " is attained, at which no
amount of increased exciting power is capable of producing any

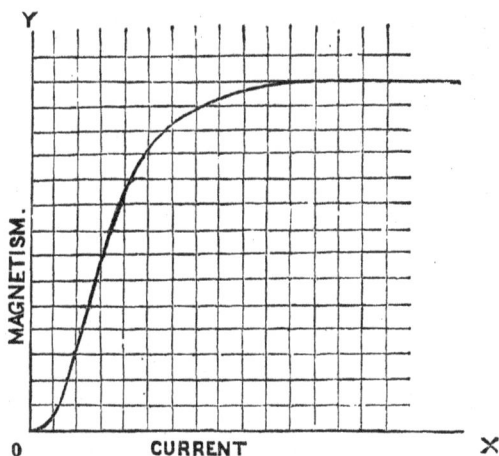

Fig. 43.—Magnetic Curve.

practical addition to the number of lines passing through the
magnet. "

This behaviour of soft iron under the exciting power of a
helical current is graphically represented in Fig. 43 by a
curve in which the strength of the exciting current is taken in
any convenient units along the line O X, and the strength of
the magnetisation along the line O Y.

It will be found when magnetising a soft iron bar of any
given cross-section that the nearer the poles are brought together the less will be the exciting power necessary to induce a
given degree of magnetisation or magnetic induction, and that

upon the complete closing of the iron circuit this reduction in the exciting power becomes enormously marked. It is thus evident that the air offers very much greater resistance to the passage of these lines of magnetic stress than does the iron. This consideration is a very important one in the construction of dynamo machines, which are in consequence always designed in such a way as to make the air gaps in the magnetic circuit as short as possible.

Electro-Magnetic Induction.—In 1820 Oerstedt discovered that a conductor conveying a current of electricity had power to control a magnetic needle. Oerstedt's experiment was merely to hold a wire, through which a current was flowing, near to a pivoted compass needle. In all cases the needle strove to place itself at right angles to the conductor, and in such a way that the north-seeking pole always turned towards the left hand, as would be viewed by a man swimming in the direction of the current with his face turned towards the needle. If in re-repeating this experiment a second wire, forming part of an independent closed circuit containing a galvanometer, be placed over the deflected needle parallel with the first circuit, which is then suddenly broken, the needle will swing back into its original position, and in so doing will generate a transient current in the newly-introduced circuit. This induced current will be in a direction opposite to that which originally caused the deflection of the magnet, and will invariably exert a quantity of energy identical with that absorbed during the original deflection, less the energy absorbed in heating the primary circuit. In this development of Oerstedt's experiment we have the phenomenon of electro-magnetic induction, the laws of which were first discovered and stated by Faraday in 1839.

Faraday demonstrated that the current-producing, or electromotive force, in a circuit moving across the lines of a magnetic field was at every moment proportional to the *rate* at which these lines were cut by the circuit. As an example of this law we will take a horseshoe electro-magnet through which a definite magnetic induction, conveniently expressed by Faraday as a definite number of lines of force, is passing. Let a single loop of wire be placed over the gap, as shown in Fig. 44. On moving the loop to one side of the gap it will cut through all the lines which cross the pole, and an electromotive force will be set up in it which will be the greater as the interval of time in which the lines are cut is diminished. If a coil consisting of n turns be substituted for the single loop, the electromotive force set up in it will be n times as great as before, and a similar augmentation may be effected by increasing the initial number of lines passing through the coil. The introduction of an iron core within the moving coil materially strengthens the initial field by reducing the magnetic resistance of the air-gap between the poles, and so considerably increases the electromotive force induced in the coil during the period of

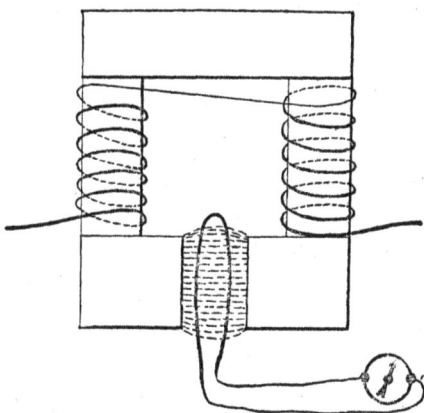

FIG. 44.—Lines of Magnetic Force.

its motion through the lines of force. If, instead of starting from the position in which the coil embraces all the lines, and then removing it from the neighbourhood of the field, we were to commence from a point without the field, and move it into the position in which it embraced the maximum number of lines, a current, resulting from an electromotive force set up in the coil, would flow as before, but in a contrary direction to that induced by the previous operation.

In making a complete phase or cycle, in which the coil enters the field from one side and leaves it at the other, passing through the point of maximum magnetic density, or, as it is generally

called, the point of maximum magnetic induction, the electromotive force generated in the circuit of which the coil is part
changes its direction at the point of maximum induction.   It
must therefore pass through zero at that point.   The actual
value of the electromotive force all through the moments of
time between the zero of entering the field and the zero of
maximum induction, and the zero of maximum induction and
the zero of leaving the field, is variable, and depends, as Faraday
stated, upon the *rate* at which the lines are cut by the circuits
at these moments, or what is in reality the same thing, the
rate at which the lines are included by or excluded from the
coil.   The identity of the above forms of stating the conditions of magneto-electric induction is not always obvious.
For instance, if we take the case of a magnet having very large
flat pole-pieces over which the magnetic density is practically
uniform, we may then consider the greater part of the field
crossing the air-gap to be also uniform, and if a very small coil
were moved across the lines of this field there would be no
change in the number of lines passing through it during the
major part of its journey ; there would, therefore, be no *rate* of
inclusion or exclusion of lines during this period and no electromotive force.   Still, it might be argued that the lines were
being cut by the circuit, and this is true, only since both the
forward and hinder parts of the coil are cutting the lines simultaneously and at the same rate, equal electromotive forces are
generated in both these portions, and, meeting together,
neutralise one another.   In the example we are taking, the
fluctuations of electromotive force would probably be of a very
complicated character, being influenced primarily by the shape
of the coil together with the shape and distribution of the field,
and secondarily by certain reactions due to the corresponding
fluctuations in the induced current.

It is important to observe that what is primarily induced in
a circuit by the increase or decrease in the number of magnetic
lines embraced by it is electromotive force.   The flow of a current must necessarily follow the generation of such a force in a
closed circuit, but it is a secondary effect, and may be reduced
indefinitely by increasing the resistance of the circuit.

When the current does flow, however, the coil is for the time
being converted into an electro-magnet, and, if provided with an

iron core to assist the passage of magnetic lines, into perhaps a very strong electro-magnet.

It was stated by Lenz, as a general law, that in all cases of magneto-electric induction the induced current was in a direction such as to offer opposition to the operation which caused it. We may, therefore, test the effect of this self-magnetisation of the coil by the application of Lenz's law.

During the approach of the coil to the position of maximum magnetic induction its faces must be so magnetised as to be repelled, and during its recession in such a way as to be attracted by the adjacent poles of the magnet. The polarity of the faces of the coil is therefore changed at the moment when it begins to exclude the lines after passing through zero E.M.F. at the period of maximum induction. Thus, during the approach, when the effect of the operation is to include more lines within the embrace of the coil, the effect of its own current is to generate lines in the contrary sense, and so to neutralise some of the lines of the initial field. During the recession, when the operation is to exclude the lines, the effect of the reversed current is to add more lines to the initial field.

The effect of current induction is thus not only to reduce the net value of the induced electromotive force, but also to delay its rise and to prolong its fall. The above reaction of a current on the electromotive force is called self-induction. It is often convenient to speak of the electromotive force or of the magnetic field as it would be without the disturbing influences of self-induction, and it is then usual to speak of the *impressed* electromotive force and the *impressed* magnetisation.

The first attempts to obtain currents of electricity by electro-magnetic induction were made on the lines of the elementary experiment which we have been discussing. Two coils were mounted upon a spindle, and were made to rotate before the poles of a steel horseshoe magnet. This form of apparatus furnished currents continually alternating in direction, and is still retained for medical purposes, and also, when separately excited electro-magnets are substituted for the permanent steel magnet, to generate alternating currents for electric lighting on a large scale at the present day.

For many purposes, however, and notably for all electrolytic processes, an alternating current is useless, and some arrangement must be made in order to rectify the alternations in the external circuit. In the elementary form of machine above described a very simple commuting device was adopted for this purpose. In the original form of the machine the free ends of the rotating coils were connected to two independently insulated rings of metal rotating with the coils. From these rings the alternating current could be collected by stationary springs or brushes making a rubbing contact on the rings. The free ends of the external circuit wire were then connected to these contact brushes.

In order to correct the alternating character of the current from brush to brush in the external circuit, a single ring

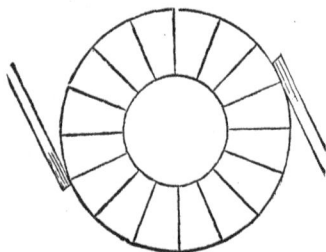

FIG. 45.—Commutator and Brushes.

divided across its diameter was substituted for the two complete rings, and the free ends of the rotating coils were joined one to each half of the divided ring. The position of the fixed springs was now so adjusted as to change their contacts with the two separate portions of the rings at the moment when the direction of the induced current was reversed in the rotating coils. The effect of this arrangement was to furnish in the external circuit a current pulsating in waves between zero and its maximum value, but always in one direction. This form of direct current, although capable of effecting electro-chemical decomposition, was still very objectionable for many purposes on the ground of its pulsating character, and a much nearer approach to the truly continuous currents furnished by primary batteries was needed.

In 1864 Pacinotti described a machine in which a number of small bobbins or coils were wound side by side upon a circular iron ring, which was mounted upon a vertical spindle, and was capable of rotation between the poles of a horseshoe electromagnet. The iron ring thus formed an armature or keeper to the magnet, whose poles were very near the rim.

This circumferential rim was provided with radial projections of iron, which served to define the angular spaces in which the coils were wound, and were also of value in reducing the airgap between the poles of the electro-magnet and the surface of the iron ring. The free ends of the coils were so joined as to form a continuous closed winding in one direction round the

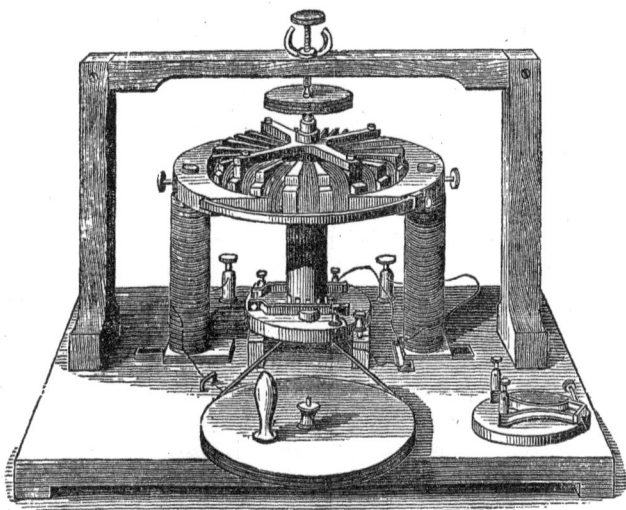

FIG. 46.—Pacinotti's Machine.

ring. At every junction between adjacent coils an electrical attachment was made to a strip of metal which was fixed longitudinally on a wooden cylinder; there were thus the same number of metal strips as there were coils or sections in the armature. The width of these strips was so chosen that although perfectly isolated from each other, the space between adjacent strips was very small. Two contact springs were arranged to press upon opposite points of a diameter on this cylindrical collector, the diameter being chosen so that contact was made by the springs to any pair of junctions on the

L

armature winding at the moment when these were midway between the true poles of the magnet. Fig. 46 indicates the general arrangement of the machine.

The presence of the iron ring between the poles of the magnet induces the lines of force, which would naturally pass in the most direct manner from pole to pole, to complete their course through the coils by which the ring is overwound. To apprehend the action of the machine, we will first consider the revolution of a single coil from one magnet pole to the other. In its position immediately over one pole the lines of force diverge through it on either side, and, passing through the iron armature core, enter the magnet at the other pole. The impressed lines of force undergo no disturbance, nor do they take part in the movement of the ring during its revolution. The effect, then, of turning a coil on the armature from a position of rest over one pole, through 180 degrees, or half a revolution, to the other pole, is similar to that which followed the operation of moving a coil through a straight-line magnetic field from a point at which the magnetic induction was zero through the point of maximum induction and out again to zero. In this latter instance we saw that the direction of the induced E.M.F. and current was reversed on passing through the point of maximum magnetic induction. Thus in Pacinotti's ring the reversal occurs in each coil as it passes through the space midway between the poles. In working such a machine, however, the coil does not start from a position of rest above one pole of the magnet, but, on the contrary, this point in its revolution is that in which the rate of change in the number of lines included in its embrace is at its maximum, the rate of change being zero midway between the poles where the induced E.M.F. passes through zero in its reversal. The E.M.F. is therefore in one direction during the half revolution of the coil, on one side between the neutral zones, and in the reverse direction during the completion of its revolution on the other side, and attains its maximum values as the coil passes the poles.

If now we consider all the coils on one side the neutral diameter simultaneously, we shall perceive that at any and every instant an induced E.M.F. in the same direction, but of different values, exists in each one of them, and that these

E.M.F.s are summed owing to the fact of the coils being joined in series to form one continuous bobbin. At the same time a similar summation of the E.M.F.s is being continually effected in an opposite direction within the coils on the other side of the neutral line. These two equal and opposite forces collide at the neutral zones, and would there neutralise each other were it not for the connections to the external circuit which are made by the contact of the brushes on opposite strips of the collector on that diameter. As it is, a current is sent through the external circuit between those two points of contact, which is the more nearly continuous as the subdivision of the armature into sections is increased, each section being connected to a separate strip on the collector.

This beautiful machine was not recognised by Pacinotti as a generator of current electricity, but was called by him an electric motor. In 1870, however, the machine was re-invented by Gramme for the purpose of generating current by the application of mechanical power. Used as a motor, mechanical power was developed by the rotating ring upon the application of electrical power in the form of current from a large battery. Viewed in this way, it is easy to see that all the coils on either side of the neutral diameter under the influence of magnetic induction resulting from the flow of a current through them would be repelled from one pole of the electro-magnet and attracted by the other, and that, being rigidly connected together by the body of the ring, their respective mechanical moments would be summed and would result in a definite tortional couple about the axis. Thus the dynamo machine is entirely reversible, and may be used with equal advantage as a motor.

In 1872, von Alteneck, improving upon an original form of machine invented by Dr. Werner Siemens, produced the so-called Siemens drum winding, in which the coils are wound longitudinally over an iron drum core, and connected together in series and severally to the segments of a col-lector, as in Pacinotti's ring. The principle of this form of machine is identical with that of Pacinotti. In both, the E.M.F.s generated in all the conductors on either side of the neutral line are summed and are directed in parallel through the external circuit. In recent years the design and construc-

L 2

tion of dynamo machinery has reached such perfection that this form of apparatus for the conversion of energy may be truly instanced as the most efficient in existence. The best dynamos to-day are capable of converting from 90 to 95 per cent. of the mechanical energy with which they are supplied into its equivalent useful electrical energy.

Such economy in working must naturally demand the almost total absence of local currents in any part of the machine. By far the greatest source of wasteful heating in the dynamos of former years could be traced to the imperfect lamination of the armature cores. When any solid mass of metal rotates in a strong magnetic field local currents are generated in it by the same mechanism as that which induces the main current in the coils of wire wound upon it. The outer portions of the mass of metal which are perpendicular to the lines of force may be looked upon as closed coils cutting the lines in their revolution, and thus becoming the seat of an E.M.F., sending currents eddying round and round in the mass. The heat which might in this way be produced in iron revolving in a strong field is enormous and quite sufficient to destroy the insulation of any wire wound upon its surface. Several devices for overcoming these inherent faults were devised by unscientific makers, such as directing a cold blast upon the armature, making the armature hollow, and providing for a cold water circulation through it, &c. These remedies only removed the worst effects of the fault, but in no way touched the cause thereof.

The only effectual way of avoiding this objectionable feature is to build up the armature core of very thin iron plates or wire, so as to ensure electrical discontinuity in directions perpendicular to the lines of force. A further condition of high efficiency is in working with a very strong magnetic field. It is obvious that the number of coils necessary to produce a certain E.M.F. will be diminished as the density of the field is increased, and this consideration lends a still greater import to that of efficient lamination, since the demand for effective lamination increases with the strength of the field. The same remarks apply to the speed of driving. An increased speed reduces the amount of copper on the machine, but it increases the chances of local heating.

PRACTICAL DIVISION.

SECTION G.

ESTABLISHING AND WORKING AN ELECTROLYTIC COPPER REFINERY.

THE chief disadvantages of the electrolytic process of refining copper are the great cost of the plant, the continuous period of time during which a large stock of metal remains unproductive of interest, the constant attention required to be paid to the process, and the comparatively large amount of covered space necessary. But notwithstanding these drawbacks, the method is rapidly extending for the refining of argentiferous and auriferous copper, because it enables the precious metals in the ore to be completely recovered, whilst by the ordinary fusion process they pass into the refined metal and are partly lost; the copper also resulting from the electrolytic process may be obtained extremely pure and of high conductivity for electrical purposes, and commands a high price.

In planning an electrolytic copper refinery, some of the first questions to be settled are the probable rate of output of refined metal, the quality of the raw copper, and the cost of the mechanical power, because upon these depend the magnitude of the plant and of all the arrangements. As the proportional amounts of horse-power, of space employed, and the number, arrangement, and size of vats, for a given daily output, differ considerably in different works, only a very crude outline of a general plan can be given (for a brief outline, *see* p. 224).

Amount of Space.—A comparatively large covered space is necessary for the refining of a moderate amount of metal. The room occupied at the North Dutch Refinery, Hamburg, for refining "330 tons a year," is stated to be about "660 square metres." That of a plant of 40 vats, worked by a "C18" dynamo of Siemens and Halske's, at Oker, and depositing "about 350 kilogrammes daily," or 125 tons per annum, "occupies 80 square metres," and at Marseilles "300 square metres" are required to refine 89 tons yearly. At the "Bridgeport Company's" Works, Connecticut, the depositing room is 100 × 120 feet, and is only one-half occupied by vats, in which "one million pounds of copper per month," = 110 tons per week, is deposited. At Stolberg, to deposit "10 to 12 cwts. a day," the surface allowed is "324 square metres." In some works—for instance, those at Casarza, near Sestri Levante, Italy—the dynamos are in the same room with the vats. In these two latter cases there are various additional chemical and other processes, which require much extra space (*see* pp. 230-232).

Total Amount of Depositing Surface.—The first consideration is the total amount of active cathode surface necessary to yield the intended amount of good refined copper in the given time. This depends essentially upon the kind and amount of impurity in the solution (and consequently also upon the kind of raw copper which supplies the impurity). It further depends essentially upon the commercial or economic condition, the relative cost of motive-power to that of copper ; where motive-power or fuel is dear or copper is cheap, the total amount of cathode surface employed is large.

If the solution contains freely metals which, like bismuth and antimony, are easily thrown down along with the copper, the total cathode surface must be much larger, and the rate of deposition much slower, in order to deposit the given amount of copper in a pure state in the given time. The rate of deposition which enables this result to be obtained, and which is usually employed in different copper-refining works, varies from about 1 to 8 or 10 ounces per square foot in 24 hours (*see* pp. 207-209). In order to refine about 1,800lb. of copper in that period of time, by a rate of deposition of 5 ounces of copper per square foot per 24 hours, 5,760 square feet of active cathode surface would be required. The more impure the solution, especially as regards the above metals, the less must be the rate of deposition per square foot of cathode surface.

With regard to the cost of motive-power, some experiments of M. Gramme's which bear upon the question, confirmed the theoretical conclusion that if we increase the number of vats in series, and at the same time keep the total amount of resistance and strength of current in them constant by enlarging the surfaces of the electrodes in each, the total quantity of copper dissolved and deposited by the expenditure of the same amount of electric energy (and consequently of motive-power) increases directly as the number of vats in series. The following table shows the conditions and results of his experiments :—

Gramme's Experiments, "Third Series."

Copper anodes ; Baths in series ; Strength of current constant ; Variable surface of anodes.

Number of Experiment.	Section of Bath in Square Decimetres.	Number of Baths.	Deflection of Galvanometer.	Kilo-grammes of active Solution.	Gramme Weight of Copper Deposited.	
					Total per hour.	Per Bath.
1	8·26	3	7·5	19·8	15·75	5·25
2	16·52	5	7·5	33·0	29·00	5·80
3	33·04	7	7·5	92·4	37·38	5·34
4	49·56	9	7·5	178·2	48·00	5·33
5	66·08	11	7·5	280·4	61·6	5·60

The resistance and strength of current were kept constant, as shown by the deflection of the galvanometer, whilst the number of baths in series was increased. The speed of the dynamo and the electromotive force of the current did not vary, and the electric energy expended was unchanged. In order to keep the resistance constant, it was found necessary to increase the section of the liquid and size of the electrodes in a greater ratio than the number of vats joined in series.

The total quantity of copper deposited was directly proportional to the number of baths. From this it may be concluded that with electrodes of unlimited surface, in an unlimited number of vats in series, a constant and limited amount of current energy would deposit a comparatively unlimited amount of copper. " The results show that with soluble anodes the expenditure of energy in the act of electrolysis is *nil*." M. Thenard has made somewhat similar experiments, and obtained similar results.   In some cases, therefore, where motive-power is expensive, as in those where coal is dear and water-power is not available, a larger total surface of anode and cathode and slower rate of deposition have been employed (*see* pp. 188, 191, 208).

We must, however, remember that whilst by doubling the number of vats in series and doubling their size, we nearly double the quantity of copper deposited by the same amount of energy in the same time, we simultaneously *quadruple* the outlay in solution, copper, and vats, and soon arrive at a point at which the increased loss of interest upon that outlay balances or even exceeds the saving of cost of motive-power required to produce the electric current.

Number and General Magnitude of Vats.—Having decided upon the total amount of receiving surface, we now require to determine its mode of distribution, *i.e.*, into how many portions the cathode surface shall be divided ? into many small vats or a few large ones ? and whether the vats shall be in single series or parallel ?   So long as the total amount of cathode surface and same density of current per square foot are maintained, the same amount of electric energy (and therefore of horse-power) will deposit the same quantity of copper per day, whether the vats be few and large or many and small ; in the former case, however, the current employed has great strength with low electromotive force, and in the latter the reverse.   The number and size are largely decided by the risk of accidental short-circuiting in any one of them, and in case it occurred the number should be such that not more than 1 or 2 per cent. of the electric energy would be wasted ; in some works two sets of vats of 20 each, in others a single series of 60, and in others two series of 120 each, have been used to deposit from 1,600 to 2,000lb. of copper every 24 hours.   The two latter only of these arrangements fulfil the above condition.   Large vats and electrodes are very inconvenient both to manage and inspect (*see* also pp. 187, 191).   With a small number of large vats in series it is difficult to detect losses of current due to leakages and short circuits ; and with a large number the loss of interest upon capital is too great unless it is compensated by greater saving in cost of motive-power or fuel.

From the total amount of active cathode surface and the total
number of vats is determined the amount of surface of anodes
and cathodes in each vat, and from the latter and the distance
asunder of the electrodes the general magnitude of each vat is
arrived at. If the number decided upon is 60, and the rate of
deposition 5 ounces per square foot per 24 hours, then the total
amount of active cathode surface in each vat to give the
1,800 lbs. is $\frac{5760}{60} = 96\cdot0$ square feet ; if the number of vats is
doubled the amount of surface in each must be halved.

Electromotive Force and Strength of Current Required.—
From the number of vats in succession (or alternations in the series
if a double row of vats in parallel is used), and the amount of resistance in each, including that due to polarisation or counter-
electromotive force, the difference of potential required at the
terminals of the dynamo is found ; and from the number in
series and the weight of copper to be deposited daily in each,
the strength of current necessary is ascertained. The amount
of electromotive force allowed for the resistance in each vat or
alternation in the series varies in different works, but to provide
for the maximum of occasional and variable counter-electromotive force of voltaic polarisation, and for producing the
maximum density of current required for quick working, a
total of ·3 to 1 volt per vat is allowed in some refineries. As
an example : at Pembrey, to deposit 4,000lbs. of copper each
24 hours, in a single series of 200 vats, a dynamo yielding 350
amperes at an electromotive force of 110 volts is employed,
and is driven by about 65 indicated horse-power from a steam-
engine.

Kinds of Dynamos Employed. — In different electrolytic
metal refineries, dynamos of the following kinds are or have
been used :—At Messrs. Elliotts' (late Elkington's), Pembrey,
near Swansea, formerly Wilde's magneto, then Gramme's and
Wilde's improved machines, but recently only those of Messrs.
Chamberlain and Hookham ; at Messrs. Bolton's, Mersey Copper Works, Widnes, Siemens's and Elwell-Parker's ; at Messrs.
Vivian's, Swansea, Elmore's, Gulcher's, Crompton's, also Edison-
Hopkinson's, made by Mather and Platt, Manchester ; at

Williams, Foster and Co.'s, Swansea, Elmore's, and subsequently Crompton's; at C. Lambert and Co.'s, Swansea, Siemens and Halske's, Gulcher's, also Edison-Hopkinson's; at Messrs. Seaver and Kleiner's, Tyldesley, Lancashire, for separating aluminium from cryolite, Edison-Hopkinson's, also Siemens's; at the Cowles Syndicate Company's Works, at Milton, near Stoke-upon-Trent, Brush, made by Crompton; at M. Letrange's, and MM. Lyon Allemand's, Paris, and M. Secretan's, St. Denis, Gramme's; at M. Hilarion Roux's Works, Marseilles, Gramme's, made by Mather, Hartford, Connecticut, U.S.A.; at M. Weiller's, Angoulême, and MM. Oeschger, Mesdach and Co., Biache, Saint Waast (Pas de Calais), Gramme's; at the North Dutch Refinery, Hamburg, Gramme's; at the Stolberg Company's, Stolberg, Westphalia, Siemens and Halske's; at Messrs. Heckman's, Berlin, and Kayser and Co.'s, Moabit, near Berlin, Siemens and Halske's; at the "Kommunion Hüttenwerke," at Oker in the Hartz, Siemens's; at the Mansfeld Mining Company's, Eisleben, and Messrs. Stern and Co.'s, Oker, Wilde's; at Messrs. Schrieber and Co.'s, Burbach, near Siegen, Siemens's; at M. Andre's, Frankfort-on-the-Maine, Gramme's; at the Aluminium Company's Works, Schaffhausen, for preparing aluminium bronze by M. Heroult's process, Oerlikon dynamos; at Stattbergerhütte, near Cologne, the Koenigshütte, in Silesia, the Hüttenwerke, Witkowitz, in Moravia, Stephanshütte, in Upper Hungary, and at the Royal Hüttenwerke, Brixlegg, in the Tyrol, Siemens and Halske's; at the works of the Electro-Metallurgical Society of Turin, Ponte St. Martino, Piedmont, Oerlikon dynamos; at those of the Electro-Metallurgical Society of Genoa, Casarza, near Sestri Ponente, Italy, Siemens and Halske's; at the Pennsylvania Lead Company's Works, Pittsburgh, U.S.A., Brush dynamos; by the "American Aluminium Company of Milwaukee," Wisconsin, Gramme's, made by Mather of Hartford, Connecticut; at the St. Louis Smelting and Refining Company's Works, Cheltenham, St. Louis, Mo., Hochhausen's; at Messrs. E. Balbach's Refinery, Newark, New Jersey, Hochhausen's, made by the "Excelsior Electric Company," of Brooklyn; at the Electrolytic Copper Company's Works, Ansonia, Connecticut, Mather's, made by the "Eddy Manufacturing Company" of Hartford, Connecticut; at the "Omaha and Grant Smelting Works," Omaha, Nebraska, Hochhausen's; at the "Bridgeport

Copper Company's Works," Bridgeport, Connecticut, Mather's; at the "Cowles' Electric Smelting Works," Cleveland, Ohio, and Longport, near New York, Brush dynamos; at M. Legger's Works, Santiago, Chili, Gramme's; at M. Moebius's, Chihuahua, Mexico, Siemens and Halske's.

In a large electrolytic refinery a considerable number of dynamos are usually employed; for instance, at the Casarza establishment, there are no less than 30 of Siemens and Halske's, and at the works at Pembrey 32 of Wilde's were formerly used.   In nearly all the works, changes in the kind of dynamo used have had to be made in consequence of improvements in those machines.   A single dynamo now deposits 15 tons of copper per week at Pembrey, and one at Bridgeport Connecticut, deposits more than 30 tons per week.

Dynamos of Different Makers.—The following are some of the chief kinds which have been used for the commercial refining of metals :—

Wilde's.—A number of Wilde's separately-excited small ones of the magneto kind were originally employed at Messrs. Elkington's works at Pembrey.   The armatures of the exciter rotated 2,400 and of the dynamo 1,500 times a minute.   They gave alternate currents, which were rectified by means of a commutator.   These machines became much heated after a few hours' working, and had to be cooled by circulating cold water through them, and by switching on cool ones in their stead at intervals of time.   They did much work, but with considerable waste of energy, and are now quite out of date.

An improved machine was brought out by Wilde in 1867. It consisted of two circular sets, each of 16 soft iron field-magnets, with their free ends facing each other, each set being attached at their outer ends to a fixed soft iron ring. Between these rotated a circle of 16 bobbins of wire with soft iron cores, constituting the armature, the coils of one or two of these being used to excite the field-magnets, and the remainder connected in parallel to generate the external current which worked the vats.   It was provided with two separate commutators, one for rectifying the currents of the field-magnets, and the other for correcting those of the armature. There were five of these machines at Pembrey, and five series of vats, each of 48 in single order, each dynamo working one series, and depositing "324 kilogrammes of copper daily." This machine was a great improvement upon the previous type, and did good service, but still produced much heat ; Chamberlain and Hookham's dynamos, a description of which is given on page 167, are now used at these works. Wilde's machines have also been employed by the Mansfeld Mining Company at Eisleben, Germany, and at Messrs. Stern and Co.'s, Oker.

According to a statement of the Electric Engineering Company, of Manchester (successor to H. Wilde), his 32-magnet dynamo is capable of depositing in 138 vats in single series—each vat having 40 square feet of active cathode surface, and the same amount of anode surface—a total of 900lbs. of copper per 24 hours, with a consumption of 12 to 13 horse-power.

The *Siemens and Halske* Dynamo.—One of these, of the "C¹" type (*see* Fig. 47) series-wound, with specially thick conductors, has been in use in the Kommunion Hüttenwerke at Oker, in the Hartz, since about the year 1878, and two have been added since. The armature is a cylinder, with the Hefner-Alteneck system of winding (see *Proc.* Institution Civil Engineers, 1878, Vol. LII., p. 39, Plate 1 ; also S. P. Thompson's "Dynamo-Electric Machinery," second edition, pp. 152, 238). It has a single layer of conductors of thick bars of copper laid longitudinally upon its outer surface only, insulated by means of asbestos, and with air-spaces between them for ventilation and cooling. Each of the four rectangular limbs of the two field-magnets is formed of seven square bars of soft iron, bolted to a thick yoke of iron at the back, and has coiled upon it a single layer of seven turns of insulated thick copper conductor, each turn having a section of 13 square centimetres. The junctions of all the conductors are bolted, and soldered together with silver solder. The collecting brushes are very solidly mounted. The internal resistance is only ·00075 ohm, the electromotive force is about 3·5 volts, and the strength of current about 1,000 amperes. The machines become much heated, but are not injured by the heat, after running day and night for 10 years. Notwithstanding the dynamos are close to the vats, and the conductors are 25 square centimetres in section, the latter become sensibly warm. Each dynamo supplies 12 large vats, arranged in series, and refines about 1·2 kilogramme per hour per vat, or

FIG. 47.— The Siemens "C" Type Dynamo.

350 kilogrammes (= 771lbs.) of copper per 24 hours, with an expenditure of 5 horse-power. The total resistance of the baths and conductors is ·0035 ohm. These machines are still in use, and the commutator of the first machine is not yet worn out.

FIG. 48.—The Siemens " H C " Type Dynamo.

At the Casarza Works 30 of Siemens' "C^{18}" vertical direct-current dynamos, shunt-wound, are used for separating copper from crude matte of iron and copper pyrites. The armature of each of these rotates about 1,000 times a minute, and yields either 30 volts and 120 amperes, or 15 volts and 240 amperes, and in the latter case, with 12 vats in single series, and an external resistance of ·0625 ohm, deposits about 180lbs. of copper every 24 hours, some of the power being wasted in deoxidising persalt of iron in solution. This improved kind of dynamo is also employed at the "Koenigshütte," in Silesia, and in the "Hüttenwerke," at Witkowitz, Moravia, and is now used in preference to the "C^1" type, in the Hüttenwerke, at Oker, where it deposits in 40 vats in single series "about 350 kilogrammes (= 771lbs.) of copper every 24 hours." Messrs. Siemens and Halske's "C^1" dynamo is also used by Kayser and Co., Moabit, near Berlin, and their "C^2" machine by M. Schrieber, at Burbach, near Siegen, Prussia; others also for refining black and red copper of 90 per cent., at Stephanshütte, Upper Hungary, and for refining 300 kilogrammes of silver daily at the works of Messrs. Moebius, at Chihuahua, Mexico (see p. 241). The following is a list of the most recent shunt-wound dynamos made by Siemens Bros. and Co. for electro-deposition. Fig. 48 shows their "H C" type :—

Type of Machine.	Maximum current in amperes.	Difference of potential in volts.	Number of revolutions per minute.	Horse-power required.
C^6 C	150	6	1,200	$1\frac{1}{2}$
C^6 N	75	15	920	2
C^7	300	8	1,180	4
C^8	240	25	850	$9\frac{3}{4}$
H C^9	300	50	750	24
H C^{10}	500	60	700	47
H C^{11}	1,000	60	650	93

The *Gramme* Dynamo.—The only works in which Gramme machines are used upon a large scale for the refining of copper are those of the North Dutch Company, at Hamburg; M. Mesdach, at Biache; and M. Leggers, at Santiago, Chili. They have, in addition, been experimentally employed for separating or refining metals, by M. Letrange in Paris, M. Roux at Marseilles, M. Secretan at St. Denis, M. Weiller at Angoulême, M. Heroult at Schaffhausen, and others.

At the North Dutch Refinery, Hamburg, there are six of the "No. 1" type (see Fig. 49), working at a maximum speed of 1,500 revolutions a minute, and giving a maximum current of 300 amperes at 27 volts. The motive energy is derived from a "40 horse-power steam-engine." When used for refining copper they are

provided with a special armature of cylindrical shape, having a
single layer of thick copper rod conductors upon its outer surface
only. "Two of these dynamos, connected in tension, operate
upon two sets of vats of 120 each, each set being in single
series, joined in tension, and deposit a total of 900 kilogrammes
in 24 hours ; the mechanical energy expended being 12 horse-
power, which equals 80,000 kilogrammetres per kilogramme
of copper deposited."

FIG. 49.—The Gramme "No. 1" Type Dynamo.

There is also used at those works a larger and much more
powerful machine of the same kind, constructed specially for
Dr. Wohlwill in the year 1873 (Fig. 50). It is 1·5 metre long,
1 metre high, ·75 metre wide, and weighs about 2,500 kilogrammes, of which 735 are copper, and the remainder iron. It
has four single cylindrical horizontal iron bars of 12 centimetres
diameter, bolted to the massive iron end-plates of the machine ;
these form eight electro-magnets, each 41 centimetres long.
On each of these magnets is wound, in 32 turns, a ribbon of

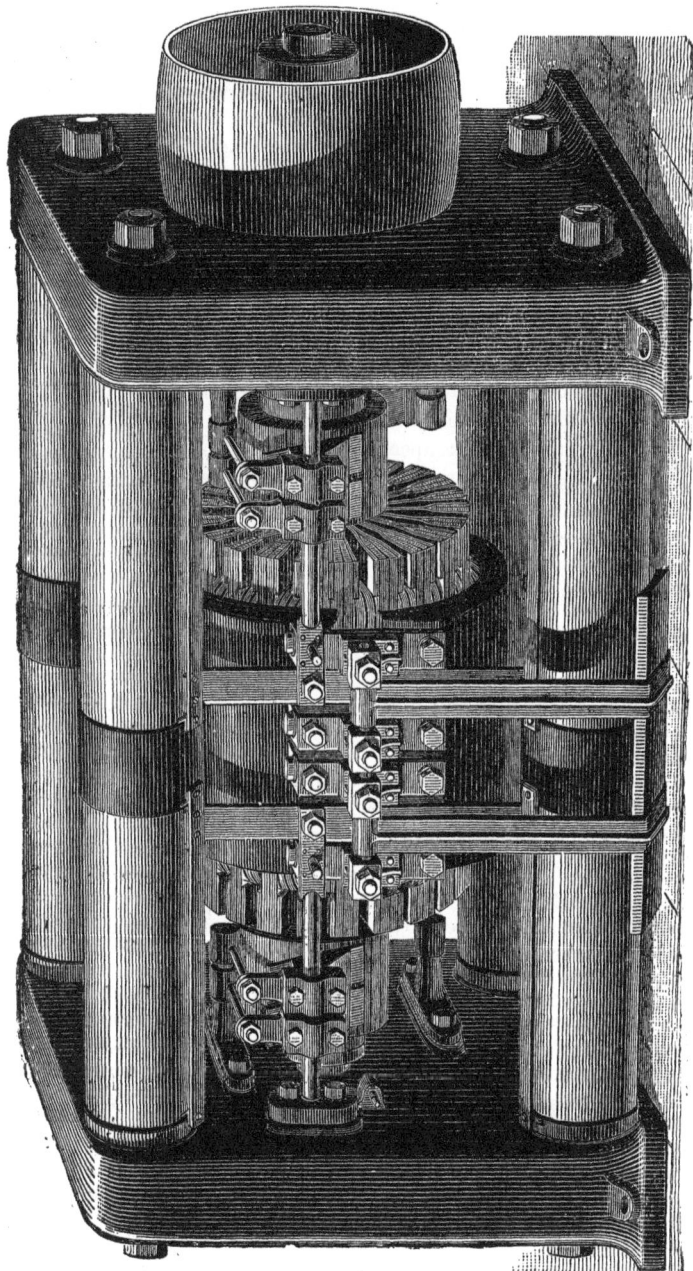

Fig. 50.—The Wohlwill Gramme Dynamo.

sheet copper, 1·1 millimetre thick, and of the same width as the length of the magnet, the eight coils being connected in series with the main circuit, and offering a total resistance of ·00142 ohm. The "ring" armature is in the form of a short cylinder. The conductors upon it are divided into 40 sections or partial coils, each section being composed of seven strips of copper, 10 millimetres wide and 3 millimetres thick. The machine has two collectors, with brushes, one at each end of the armature, each collector having 20 sections. Twenty of the partial coils are connected to the right-hand collector, and the other 20 to the left-hand one. The total resistance of the armature conductors when connected in series is ·0004 ohm, and when in parallel ·0001 ohm; in the former case, with a speed of 500 revolutions a minute, the electromotive force is 8 volts and strength of current 1,500 amperes, and in the latter 4 volts and 3,000 amperes. It supplies 40 baths, in two parallel series of 20 each, and deposits 800 kilogrammes of copper every 24 hours, with a consumption of 16 horse-power. This result is inferior to that obtained with the smaller dynamos.

It is stated that at this refinery 500 tons of copper are electrolytically refined each year, and that the deposited copper is exceptionally pure. (United States Geological Survey, "Mineral Resources of the United States," by A. Williams, 1883, pp. 225, 644, and 1884, p. 369, published at the Government Printing Office, Washington.)

At M. Hilarion Roux's works, Marseilles, "a Gramme 'No. 1' dynamo, with its armature revolving 850 times a minute, yielded a current of 300 amperes and an electromotive force of 8 volts, and with 40 vats deposited 10·4 kilogrammes of copper per hour, with an expenditure of 5 horse-power and a daily consumption of 240 kilogrammes of coal." Several Gramme machines were formerly used at Pembrey.

The *Brush* Dynamo.—Used by the Pennsylvania Lead Company, Pittsburgh, U.S.A., for refining lead; also by Messrs. Cowles and Co. for separating aluminium, silicon, &c., in the form of alloys. An enormous one has recently been employed for this purpose at Longport, near New York. It is 15 feet long, 5 feet high, and 4 feet wide, and weighs about 9¾ tons. (Fig. 51). It is compound-wound, and yields direct currents. It has eight field-magnets, each with a cylindrical core of cast iron 16in.

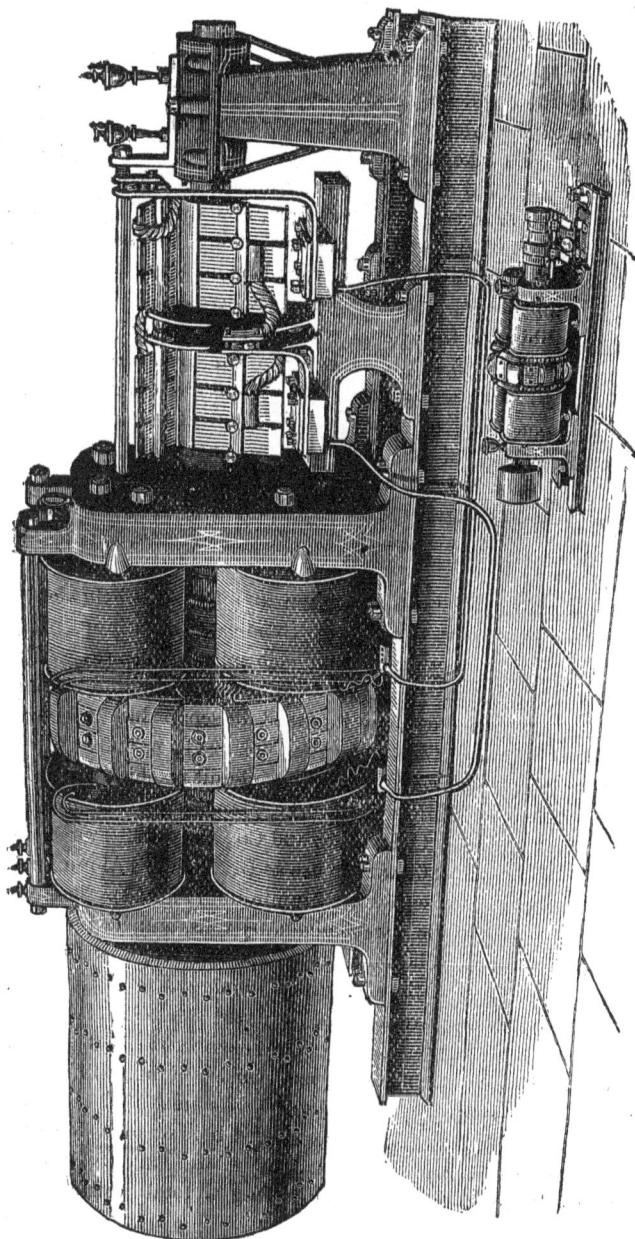

FIG. 51.—The Brush Dynamo.

M 2

long and 11in. diameter, and wound with 30 layers of 102 turns
each of single copper wire ·134in. (= 3·404 millimetres) diameter ; all the eight wires are coupled in multiple arc, and
have, when thus combined, a total resistance of 1 ohm when cold.
The magnet coils take a current of 80 amperes, or about 2·5
per cent. of the total current.　The armature is 42in. diameter,
contains 1,600 pounds of wrought iron, and has 16 bobbins ;
each bobbin has 21 turns of copper wire, 65ft. long and ·35in.
diameter, in two parallel strands, wound upon it ; the bobbins
are all in multiple arc.　Sixteen copper bars convey the currents
from the bobbins to the commutators.　The weight of copper
on the magnets is 5,424lb., and upon the armature 825lb.
It is stated to yield, at a speed of 450 revolutions a minute, and
with an expenditure of nearly 400 indicated horse-power, an
electromotive force of 80 volts, and a strength of current of
3,200 amperes, or 249,000 watts of energy, and to be capable of
yielding 300,000 watts.　It is driven by two 30in. turbines,
and absorbs 355 horse-power.　(See *The Electrician*, Oct. 15th,
1886.)

More recently, Messrs. Crompton and Co., of Chelmsford,
have constructed for the Cowles Syndicate, at Milton, near
Stoke-upon-Trent, another of these colossal machines (Fig. 52),
yielding 5,000 amperes, having a guaranteed minimum working
capacity of 300,000 watts, and intended for separating the above
refractory metals.　(*The Electrician*, Vol. XXI., p. 590.)

The *Edison-Hopkinson* Dynamo.—A 50-unit machine made by
Mather and Platt, Salford Ironworks, Manchester.　In use by
Messrs. Vivian and by C. Lambert and Co., Swansea.　Two of 90
horse-power and one of 60 were in use at Dr. Kleiner's Aluminium
Works, Tyldesley, for separating aluminium (Fig. 53).　The bar
armature type machine weighs about $5\frac{1}{4}$ tons, and its magnets
and pole-pieces are solid forgings of specially high quality of
soft iron.　The core of the armature is composed of about 1,000
discs of very thin charcoal iron, insulated from each other by
thin sheets of paraffined paper.　The machine is shunt-wound ;
the conductors on the field-magnets consist of 520 pounds of
copper wire, and have a resistance of 3·74 ohms.　Those on the
armature are formed of 74 wedge-shaped bars of drawn copper,
each of ·338 square inch sectional area, insulated from each
other by " a special tape," and having a total resistance of

0·016 ohm at 13·5°C. The commutator is formed of copper bars, insulated with mica, and has five separately adjustable

FIG. 53.—The Edison-Hopkinson Dynamo.

spring brushes on each side. At a speed of 400 revolutions a minute the machine gives 50 volts and 1,000 amperes. The electrical efficiency of the machine is between 95 and 96 per

cent., and the commercial efficiency "between 93 and 94 per
cent.," meaning by the term " commercial efficiency " the ratio
of the electrical power in the external circuit available for useful work, to the mechanical power absorbed by the machine.
    The *Chamberlain and Hookham* Dynamo (Fig. 54).—A num-
ber of these machines are used by Messrs. Elliott and Co., Pembrey. "The 30-unit machine (or 25-unit nominal) weighs
25 cwts.; it is shunt-wound and its armature and pole-pieces
are magnetically well isolated from the iron framework. The
field magnets have a cross-sectional area of 42 square inches,
with yokes and pole-pieces of cast iron; they are wound with
2,856 turns of ·109in. copper wire, having a total resistance of
8 ohms when hot; the amount of current shunted through
this wire is 7·13 amperes. The armature is cylindrical, 13in.
long and 10in. diameter, and built up of notched discs of very
thin sheet iron and paraffined paper. It has 35 longitudinal
slots and projections (like those of the Pacinotti ring), each slot
being slightly oblique to the axis, in order to diminish singing
vibrations, and containing 12 insulated wires in parallel, having
a total resistance of ·003 ohm when hot. The entire length of
wire on it is about 53 yards, or 32ins. per volt generated. At a
speed of rotation of 900 per minute, the machine yields a current of 450 amperes at 57 volts. The total amount of energy
converted is 26,635 watts, equal to 9·5 watts per pound weight
of material in the machine; 610 watts are absorbed in the armature conductors, thus giving a loss of 2·28 per cent.; there is
also a loss of 373 watts, or 1·4 per cent., in the wires of the
magnets. The electrical efficiency is 96·3 per cent., and the
commercial efficiency is about 94 per cent. The total weight
of the machine is in the proportion of 82 pounds per electrical
horse-power developed in the external circuit."
    "The 60-unit machine for electro-depositing has an armature
25in. long and 10in. diameter. At 820 revolutions a minute,
it yields 110 volts and 350 amperes, and deposits 4,000 pounds
of copper per day of 24 hours, or 20 pounds per vat in 200 vats.
The 30-unit machine at 900 revolutions gives 55 volts and
350 amperes, and deposits 2,000 pounds of copper per day of
24 hours, or 20 pounds per vat in 100 vats." One of the 60-
unit machines is stated to have deposited 18 tons per week.
These machines are only worked up to 70 per cent. of their full

FIG. 54.—The Chamberlain and Hookham Dynamo.

capacity, in order to obviate undue heating.   One of the former
and five of the latter, including a spare one, are used at a single
works.

The *Elwell-Parker* Dynamo.—The largest machine of this
type hitherto made is a 75-unit machine, with four poles and
two horizontal wrought-iron magnets (Fig. 55).   It is plain
shunt-wound, and has an armature of the drum type 22in.
diameter and 20in. long, having a core of thin sheet-iron discs
mounted upon the shaft, and 80 parallel copper wires on its
surface, each of ·2 square inch area of section, and having a
total active length of 1,600 inches, or about 47 per cent. of the
total length.   The resistance of these wires is ·0008 ohm when
cold, and of those upon the magnets 1·25 ohm.   The loss of
energy in the armature is 1,800 watts, and in the magnets
2,000 watts.   It is stated by the makers to yield a current of
1,500 amperes, and a difference of potential of 50 volts, when
revolving at a peripheral speed of 2,500ft., or 500 times a
minute, by an expenditure of 94 approximate belt horse-power,
and to have an electrical efficiency of 95·1 per cent., and a
commercial efficiency of over 90 per cent., without sparking at
full load.   The density of current in the armature wire with
this load is equal to 1,880 amperes per square inch sectional
area of wire.   Its general dimensions are, 8ft. 2in. long, 6ft.
wide, and 3ft. 8in. high.   Its weight is about six tons, equal to
about 130lbs. per horse-power of electrical energy developed in
the outer circuit.   Three of them are in continual use, running day and night, usually about 160 hours per week, at the
Mersey Copper Works, Widnes, Lancashire.   (*The Electrician,*
Vol. XXI., p. 183.   Esson's "Dynamo-Electric Machines," 1887,
p. 288.)   It should deposit 11·44 tons of copper per 156 hours,
in a single series of sixty vats, when running at the above
speed.

The 50-unit machine is a two-pole one, with vertical electro-
magnet, and a drum armature, giving "50 volts and 1,000
amperes."   "The resistance of its armature coils is ·0023 ohm,
and of the magnet coils 3·72 ohms.   The loss of energy in its
armature is 2,300 watts, and in the magnet 675 watts.   Its
electrical efficiency is 94·3 per cent., and its commercial
efficiency over 90 per cent.; length 9ft., width 2ft. 10in., height
3ft. 4in."

FIG. 55.—The Elwell-Parker Dynamo.

The *Gülcher* Dynamo.—Several of these are used at C. Lambert and Co.'s., and Messrs. Vivian's works, Swansea. The machine (Fig. 56) is composed of two sets each of four soft-iron field-magnets, each set being arranged in a circle and fixed at their outer ends to a cast-iron end-plate forming the yoke, with their inner ends facing each other, and having hollow box-shaped pole-pieces of cast iron fixed upon their ends, within which the armature revolves. Each alternate bar has opposite polarity, and each opposing magnet similar poles, and they are all shunt-wound. The armature has the appearance of a flat ring or disc, and is formed of a gun-metal wheel, having a ∟-shaped rim, in the angles of which are wound two continuous ribbons of soft iron insulated by asbestos paper. This composite ring is turned true and its angles rounded, and the insulated conductor wound at right angles upon it in one or two layers of continuous sectional coil (as in a Gramme ring) covering its entire surface. It thus forms a kind of fly-wheel revolving between the two circles of magnets, and within their hollow pole-pieces, and from its form, construction, &c., is well ventilated. Its axle has massive bearings. The collector, and its connections with the armature coil, are like those in a Gramme machine ; it is very substantially formed of hard-drawn copper bars, insulated with talc.

4-Pole Gülcher Dynamos.

Units.	Amperes.	Volts.	Vats in series.	Square feet of cathode surface per vat.	Brake at pulley. Horse-power.	Pounds of copper deposited per hour.	Revolutions per minute.
2	200	10	20	20	3½	10	1,500
3	300	10	20	30	5	15	1,200
5	500	10	20	50	8½	25	1,000
10	500	20	40	50	17	50	800
12	500	24	48	50	20	60	750
15	500	30	60	50	25	75	600
20	500	40	80	50	35	100	550
25	500	50	100	50	40	125	450

"These machines are made of two types, viz., No. 4, yielding a current of 700 to 800 amperes at a difference of potential of 20 volts, with an electrical efficiency of 93 per cent., and commercial efficiency of 87 per cent., and No. 6, giving 500 to 600

Fig. 56.—The Gülcher Dynamo.

amperes at 50 volts, with an electrical efficiency of 94 per cent., and commercial 89 per cent. The latter size is designed to work 100 vats in single series, each vat being 3½ft. long, 3ft. wide, and 3ft. deep, and containing 50 square feet of cathode surface, equal to a density of current of 10 amperes per square foot of depositing surface." The foregoing table is the maker's list of these machines; the five larger sizes only are used for refining, and the smaller ones for plating.

The *Oerlikon* Dynamo (Fig. 57).—Made at Oerlikon, Switzerland. This is a shunt-wound machine, and has a capacity of

Fig. 57.—The Oerlikon Dynamo.

50,000 watts; five similar ones are employed by the Società Elettro-Metallurgica of Turin in their refinery at Ponte St. Martino, Piedmont. Its total weight is 14,700lbs., its electrical efficiency 95 per cent., and it yields a current of 400 amperes at a potential of 120 volts. The resistance of the field-magnet coils is 11·1 ohms, and of the armature coils ·0075 ohm. A current of 10·36 amperes flows through the coils of the field-magnets. After 14 hours continuous running with full load, the following fixed temperatures were attained :— Armature, 140° Fahr.; field-magnets, 85° Fahr. (*The Electrician*, 1887, Vol. XIX., pp. 291, 306). Two more, each yielding 6,000 amperes at 16 volts, are constructed for separating aluminium at Schaffhausen by M. Heroult's process (p. 258).

The *Hochhausen* Dynamo (Fig. 58).—This dynamo has a closed-coil armature in the form of an elongated ring, made of four separate curved frames of iron, carrying the previously

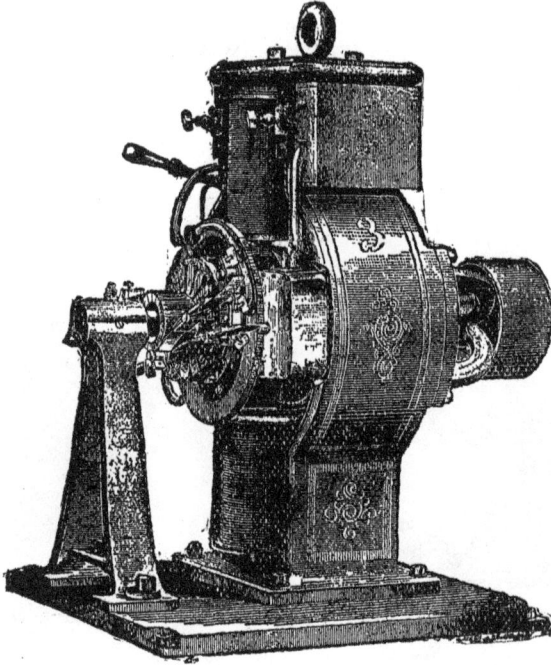

FIG. 58.—The Hochhausen Dynamo.

wound coils, and bolted to massive end-plates. It nas two vertical field-magnets, one above and the other below the armature, the upper end of the former being bolted to and supported by two vertical curved frames of iron, one on each side, fixed to the bed-plate of the machine. The segments of the collector are very massive, with air-spaces between them, and are bolted to a thick disc of slate (Fig. 59). This dynamo is made by the "Excelsior Electric Company" of New York city, and is stated

by its inventor to have a commercial efficiency of " about 85 per
cent."

" Five of these machines, viz., three ' No. 7 ' and two
ᐧNo. 6,' are used at the works of Messrs. Edward Balbach
and Sons, Newark, New Jersey, and deposit altogether 60 tons
of copper per week, running day and night, Sundays included.
Each ' No. 7 ' dynamo deposits 2·1 tons per day in 48 vats
in series.   Each ' No. 6 ' deposits 1 ton per day, one of
them working 12 large vats and the other 60 small ones.   At
the St. Louis Smelting and Refining Company's works, Cheltenham, St. Louis, Missouri, a ' No. 7 ' deposits 14 tons of
copper per week in 48 vats ; and at the Omaha and Grant

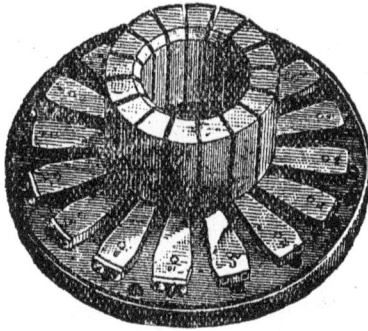

FIG. 59.—Hochhausen Collector.

Smelting Works, Omaha, Nebraska, a 'No. 6' operates 48 small
vats, and has a capacity of depositing 7 tons of copper per
week.   Each ' No. 6 ' is driven by a 25 horse-power high-speed
Westinghouse steam-engine, and each ' No. 7 ' by a 50 horse-
power engine of the same kind."

The *Mather* Dynamo (Fig. 60). — This dynamo, as con-
structed by " The Eddy Electric Manufacturing Company," of
Windsor, Connecticut, has been made up to a size of 65 horse-
power, for the purpose of refining copper, &c., and to give a high
electromotive force suitable for operating many vats in series.
Three dynamos of this kind are used by " The Electrolytic
Copper Company," of Ansonia, Connecticut.   " The Bridgeport
Copper Company," of Bridgeport, Connecticut, possess two 60

horse-power and two 45 horse-power Mather dynamos, and refine by the electrolytic process the entire product of " The Parrot Silver and Copper Company," of Butte, Montana. They employ three dynamos, each of about 40,000 watts capacity, and refine by a special electrolytic process altogether " about one million pounds of copper per month," or 112 tons per week, by means of them, and they state that they " could double this quantity for the three machines."

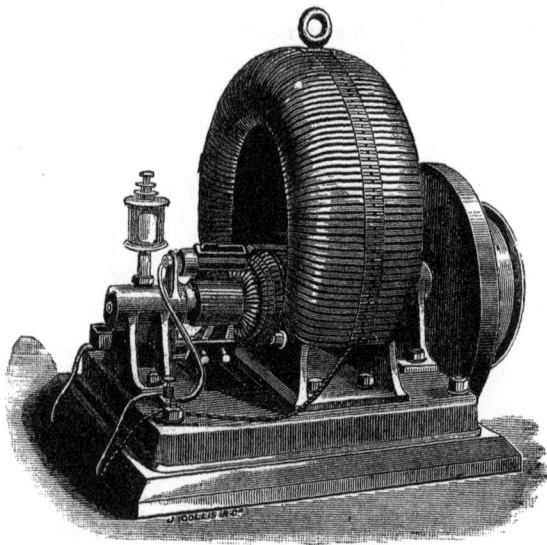

Fig. 60.—The Mather Dynamo.

The *Crompton* Dynamo.—One of these, of the " No. XL." type, has been used by Messrs. Vivian and Co., Swansea, for refining copper ; it yields an electromotive force of 10 volts, and a strength of current of 500 amperes (*The Electrician*, Vol. XVI., p. 400). It has also been employed at Messrs. Williams, Foster and Co.'s, Swansea. I have been unable to obtain further information respecting any electro-refining use of this machine.

The *Elmore* Dynamo.—A large machine of this type has been used for refining copper at the works of Messrs. Vivian and of Williams, Foster and Co., Swansea.

Source and Amount of Motive-power.—The kind of motor employed is usually a steam-engine, but in some works turbines are used. A good steam-engine only yields in the form of mechanical power to a dynamo or other receiver about 12 per cent. of the energy contained in the coals; a gas-engine, of not too large dimensions, consuming 20 cubic feet of coal-gas per horse-power per hour, is estimated to yield about 17 per cent. of the energy of combustion of the gas ; and a good turbine, worked by water, yields 80 or 85 per cent. of the energy of the water. With a Parson's turbo-generator or steam-turbine, the estimated return of the energy contained in the steam is as high as "70 per cent."

The amount of mechanical power necessary to drive a dynamo required to deposit a given amount of metal per day varies in different works, in consequence of the different proportions of waste of energy ; it also varies with different dynamos. We know that 1 horse-power is equal to 746 watts of electric energy, and that in the absence of all resistance, counter-electromotive force, and loss of current, that number of watts in the form of 746 amperes at one volt difference of potential will deposit 1·93lbs. of copper per hour in a single vat ; but in practice we know that in consequence of loss in transformation of energy, electric conduction resistance, polarisation, leakages of current, and chemical corrosion, a large allowance has to be made, amounting to at least 10 per cent., for losses in the dynamo, and a variable percentage for those in the external circuit (*see* pp. 201-204). To deposit in a single vat about 1,800lbs. of copper each 24 hours, without allowing for those losses, would require the electric energy of 28,980 watts and the mechanical energy of about 38·86 horse-power. The actual amounts of mechanical and electrical energy consumed at different works, in depositing given weights of copper, are stated in subsequent paragraphs (pp. 198, 200).

Choice of Dynamo.—In selecting a dynamo we are guided less by its capacity than by its commercial efficiency. The chief circumstances which have to be considered are :—1st, the amount of its available electrical energy in proportion to that of the mechanical power supplied to it; 2nd, the relation of its electromotive force to the total resistance and

N

counter-electromotive force to be overcome. In this country experience has led to a choice of dynamos yielding about 50 volts in preference to a lower electromotive force ; 3rd, the strength of its current to the quantity of metal to be deposited per hour, with sufficient allowance for waste ; 4th, its power of self-regulation ; 5th, the extent to which it becomes heated by long-continued running whilst doing a full measure of work ; 6th, its durability and freedom from sparking at the brushes ; and 7th, its price. Dynamos for electrolytic refining are usually constructed to order, stating the numbers of amperes and volts required by the refiner.

For the electrolytic refining of copper on a large scale a special kind of dynamo is required. It should yield a full proportion of *available* electric energy in return for the amount of mechanical power supplied to it, say not less than ninety per cent. It should be self-regulating ; this is usually effected by having powerful field-magnets, little or no resistance in the armature, by shunt-winding, and an electro-magnetic governor to regulate the supply of steam to the engine.* It should be amply large in proportion to the work to be performed, including loss by polarisation and waste of current, so as not to have to run at too great a speed. The speed of rotation of the armature should be moderate, because it has to run day and night for a week on each occasion without stopping. The armature should be well balanced and well ventilated. By long continued rapid running it is apt to swell. Whilst the conductors on the armature should be large in section in proportion to the strength of the current, so as not to generate much heat of conduction resistance, both they and the iron portions of the armature should be so divided as to be as free as possible from induction currents. It should have little or no lead at the brushes, and should not emit sparks ; the commutators and brushes are the parts which have the most wear, and which usually give the most trouble. It should not be injured by heat or work ; the hottest parts of the insulating material in it should be composed of some incombustible substance, such as asbestos or talc. The field-magnets should be very strong, and there should be just enough iron in the armature and no more,

* For a description and sketch of Willans' electric governor, see *The Electrician*, Vol. XIV., p. 208.

so that it becomes exactly saturated with magnetism when the maximum strength of current is being produced.   The electromotive force of the current should be proportionate to the resistance and maximum amount of polarisation in the external circuit, and the strength of current should be sufficient to deposit the metal at the desired rate with sufficient allowance for waste.

In different dynamos used for electro-metallurgical purposes the resistance of the conductor upon the armature varies from about ·0001 to ·01 ohm.   As the total electric energy produced by the dynamo is divided between itself and the external circuit directly in proportion to their resistances, when there is no counter-electromotive force in the external circuit the armature coils should offer very small resistance to the current.

The speed of revolution of the armatures of different dynamos used in electrolytic refining works usually varies from 400 to 1,000 revolutions or more a minute, but in some works it is greater, and with a given dynamo it is varied according to the degree of freedom of the solution from metals likely to be deposited along with the copper, and to other circumstances; if, for instance, several of the vats happen to become short-circuited, the speed should be reduced.   At large refineries a spare dynamo is usually kept in reserve.

For further particulars respecting the choice of a dynamo for electrolytic refining, the reader is referred to books specially devoted to a description of that machine.

Care of Dynamo.—The room in which it works should be kept cool, dry, and free from dust by means of doors of fine wire-gauze.   The dynamo should not be allowed to become so hot as to be uncomfortable to touch; when this occurs there is too much current produced, either by too great speed, too heavy work, or through short-circuiting or leakage of current.

On no account must any piece of metal be placed across the terminals or other parts of the dynamo, or any changes be made in the permanent connections of the circuit whilst the machine is in motion.   The coils must be kept perfectly dry.

The dynamo should be run in the proper direction and not at a greater speed, or to produce a stronger current, than that for which it is intended; the speed should also be kept uniform.

N 2

The belt or ropes with which it is driven should be equally thick and flexible throughout.

The dynamo should be kept clean, and its bearings oiled by means of a brass or copper oil-cân—not an iron one, because it may be drawn into the machine by the magnetism. The collector should be cleaned as often as is seen to be necessary from copper dust, thick oil, and dirt, and be brushed with a hard brush in the direction of the sections, and wiped with a rag slightly moistened with mineral oil till perfectly clean ; it should also be occasionally polished with fine emery cloth, and the edges of the brushes trimmed with a smooth file. During running, it should be slightly oiled by means of an oily piece of cloth, not by loose cotton waste, lest the loose fibres may get under the brushes. In a well-constructed dynamo, the commutator lasts for ten or more years.

If sparks appear at the brushes, the latter should be moved altogether backward or forward a little, and made to bear firmly upon the collector until they cease ; if this does not entirely prevent them, the commutator has probably become uneven, and requires to be made accurately true in a lathe. About once a week the brushes should be moved forward a little, to make up for their wear, and when the dynamo is not running they should be raised a little out of contact with the collector ; they must not be allowed to touch the framework of the machine.

The Depositing Room.—This should be covered, very dry, and well lighted from above, and be capable of being kept at a fairly uniform temperature, never very cold. It should have a pure atmosphere; foul air causes great trouble by damaging the electric contacts. The magnitude of the room depends upon the size and number of the vats; for sixty of average dimensions a space about 60ft. or 70ft. long and 30ft. to 40ft. wide will be large enough (*see* p. 149). It should be sufficiently lofty to allow of the most elevated of the vats standing upon a wooden stage 4ft. or 5ft. above the floor, and, if heavy plates are used, to admit of a light travelling crane being worked overhead for shifting them.

The Vats.—These require to be very carefully made ; they are usually formed of thick deal, lined with lead. In constructing the lining the edges of the lead are melted together by means of the hydrogen blow-pipe, as in making vitriol chambers. They are a frequent source of trouble, and are apt to leak both the liquid and the current; sometimes minute holes occur at the joints; in other cases the lining is injured by the sharp corners and projections of the electrodes whilst lifting them in and out; occasionally, also, sharp-pointed particles of copper stick to the lead and cause corrosion, and nodules and patches of copper form upon it electro-chemically. Some have been made of wood, lined with cement and coated inside with asphalte. Sometimes they are formed of slabs of slate bolted together, and lined with a mixture of pitch and sand. Slate is too expensive for very large vats.

Those at Casarza are " $6\frac{1}{2}$ft. long, 3ft. deep, and $1\frac{1}{4}$ft. wide ;" at Pembrey they are about 4ft. long, $3\frac{1}{2}$ft. deep, and 3ft. wide ; at Biache they are " nearly 3in. thick, 3 metres long, 1 metre deep, and ·8 metre wide." The usual dimensions in this country are about 4ft. long, 3ft. wide, and 3ft. or 4ft. deep. All the vats in a series are of the same size ; their dimensions are largely determined by facility of access to the electrodes, and readiness in shifting and removing them. They should be sufficiently wide to leave about 2in. of space on each side from the edges of the electrodes, of such a length as to allow about 2in. or 3in. between each electrode, $\frac{3}{4}$in. for the thickness of each electrode, and sufficiently deep that the lower edges of the plates do not disturb the mud. With very impure anodes, such as copper matte, the depth and length of the vats are somewhat greater because of the more copious deposit of sediment and the thicker anodes. The larger the tanks the fewer of them required, and the larger the total amount of surface of electrodes in each the greater the deficiency of deposited metal if short-circuiting occurs ; if they were very large this loss would be serious ; it is best, therefore, not to employ more than a small proportion, say, one-fiftieth to one-hundreth of the total electric energy and of depositing surface in one vat or alternation in the series (see p. 182).

Insulation of the Vats.—Everything about the outsides and bottoms of the vats should be perfectly dry; the vats insulated from each other, and from the ground, as completely as possible ; and the stage which supports them should be sufficiently high to allow the bottoms of them to be thoroughly examined from beneath. In the works at Biache, each vat is sufficiently distant from the next one in series as to allow a free passage between them. This is necessary when the vats are large.

Arrangement of the Vats.—This may be—1, in series ; 2, in parallel; or 3, combined series and parallel. In the " series " plan they are arranged in single order, their number is large, and the amount of surface of electrodes in each is small ; in the " parallel " method, the electrodes are all connected together as if they were in one vat ; and in the " combined series and parallel," their number in series is small and the amount of surface of electrodes in each alternation is large. " There is a firm near Swansea using the electrolytic process, and over 1,500 vats, arranged in various ways." " In some cases there are 100 or more in series, four in multiple arc, making 400 to 500 vats in a circuit." (U.S. Geological Survey. " Mineral Resources of the United States," by A. Williams, 1883, p. 643; published at the Government Printing Office, Washington.)

The arrangement of a number of small vats in series is preferable, because with a small number of large ones in series, and a current of low electromotive force, any increase of resistance in the vats, or any cross-circuiting, has much more serious effects than with a larger number of vats and a current of higher electromotive force ; small vats are also more convenient of access than large ones. Using vats in series is said to " steady the current." In this country the single series system is usually adopted, whilst at some works on the Continent, where motive-power is expensive, the combined series and parallel has been used. In nearly all cases, the vats are placed upon an inclined wooden platform, formed of a series of very shallow steps, to cause the liquid to flow from one vat to another cascade fashion (see Fig. 61).

In England it is usual to employ about 60 or more vats in single series, with the vats end to end, very near together, and a gangway between each single row, or between each double row, for convenience of access to the electrodes, the current going up each single row and down the next. At Pembrey, there were formerly six sets, each set consisting of 48 vats in single series and worked by a single Wilde's dynamo ; more

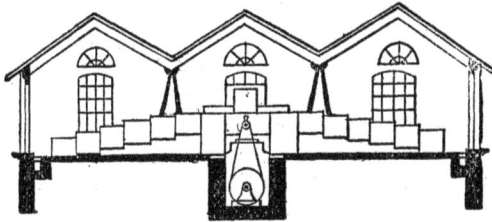

FIG. 61.—Section of Vat-Room at Casarza.

recently one set of 200 vats, placed end to end, and about 2in. apart, connected in single series, and worked by a 60-unit

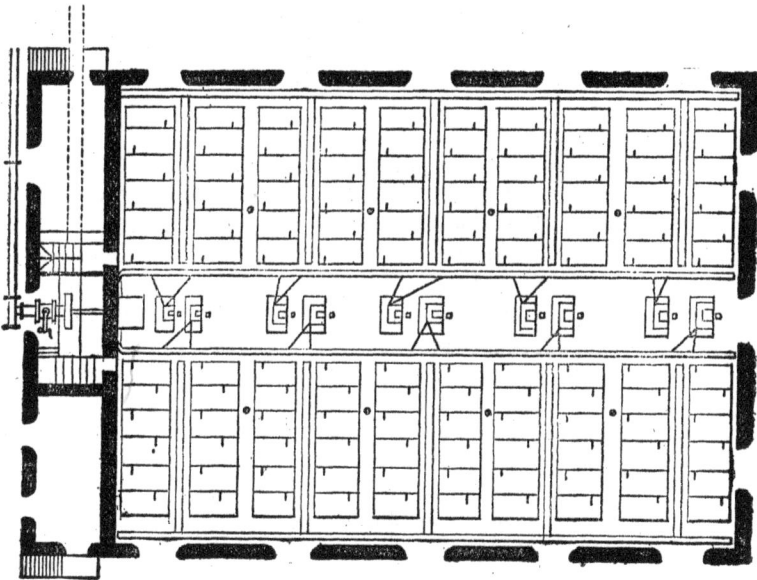

FIG. 62.—Plan of Vat-Room at Casarza.

Hookham and Chamberlain dynamo, and arranged in two double rows of 50 each ; also four sets, each consisting of 100

vats, connected in single series, arranged in two double rows
of 25 each, each set being worked by a 30-unit dynamo; and
a separate 50 vats in single series, worked by a 5-unit dynamo,
for making cathodes.   At Biache there were "40 vats in
single series."   In the North Dutch Refinery at Hamburg
"240 tanks in two series of 120 each," worked by the current
from two dynamos connected for tension ; also 40 vats in two
series of 20 each, worked by a single dynamo (see p. 160).   At
Casarza each dynamo works 12 vats in single series (see Fig. 62).
At Oker, with the "C^1" type of Siemens and Halske's dynamo
(see p. 157), there are 12 large vats in series, and with the
"C^{18}" dynamo, 40 vats in single series.   One writer (Dr. Higgs)
speaks of as many as "1,500 depositing cells" being used, by
which, with one dynamo, "as much as three tons of copper
have been deposited daily ;" also of "one set of 327, placed
109 in series and 3 in multiple arc."   The Ansonia Copper
Company work from 30 to 75 (usually from 50 to 60) vats
in single series, with each dynamo of 30,000 watts, yielding
300 amperes at 100 volts, and " are under contract to deposit
400,000lbs. of copper per month."

The Electrodes.—The sizes of anodes and cathodes are
chiefly determined by convenience in handling, and as the
anodes are thick and heavy, they are in some refineries much
smaller than the cathodes.   The number of electrodes in each
vat varies greatly in different refineries; a usual number in this
country is eight anodes and nine cathodes.   At Pembrey each
vat contained "16 anodes and 10 cathodes ;" at present there
are in each vat practically five anodes and four cathodes,
each anode being formed of four separate pieces about eight
inches wide and two feet deep, suspended edge to edge.   In
the establishment at Biache as many as 88 anodes and 69 larger
cathodes were employed in each vat, the anodes and cathodes
having about an equal total amount of active surface, and in
other works only 10 cathodes per vat.   At Casarza each vat
contains 15 anodes and 16 cathodes ; at Marseilles "115 plates
in each vat."   In some works the anodes are numerous and
narrow, about ·15 to ·175 metre wide ; whilst in others they
are ·5 metre or more.   At Biache there were 22 rows of anodes
4 in a row, and 23 of cathodes 3 in a row.   One difficulty, both

with single wide electrodes and with several narrow ones
suspended side by side, is to keep the anodes and cathodes
sufficiently parallel to each other, which is one of the chief
conditions for obtaining deposits of uniform thickness.

The average distance of the electrodes asunder varies in
different refineries from 5 to 9 centimetres; at Biache it is
7 centimetres, at Hamburg 6·3, at Pembrey 6·0, at Marseilles
5 centimetres, and it should not be less because of impurities
from the anodes getting upon the cathodes, also because of the
risk of mutual contact and short-circuiting. These effects occur,
sometimes by fragments of the anodes falling over, at other
times by rapid growth of nodules of copper upon the cathode,
especially when the current is extra dense. The general
arrangement of the electrodes is shown in the annexed sketches
(Figs. 63, 64), copied from *Industries*, Vol. II., 1887, p. 517.

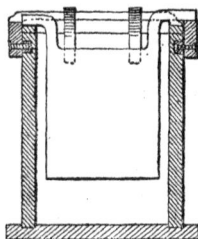

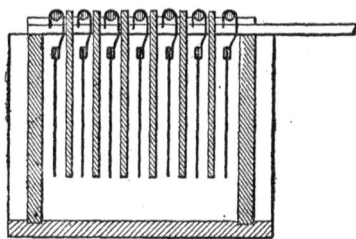

FIG. 63.—Cross-Section of Vat.   FIG. 64.—Longitudinal Section of Vat.

In some works the electrodes are suspended in a direction
across the vats, the main conductors being along the side
edges; whilst in others they are along the vats, and the conductors upon the end edges; the latter arrangement is in use
at Pembrey. At Biache the two sets of electrodes in a vat are
supported by two separate brass frames above, made in one
piece each, the positive one being over the negative, and both
well insulated from each other and from the vat. At Casarza
all the electrodes are attached to the cylindrical copper main
conductors by means of thin strips of sheet copper, with copper
wire wound upon them, and tightened by means of a screw, as
shown in Fig. 65. At those works the upper edges of the
anodes are ·8 inch above the level of the liquid, whilst the
cathodes are wholly immersed.

One inventor (E. S. Hayden, Patent No. 2,071, February 11, 1888) refines plates of metal by arranging a series of them parallel and vertically in a suitable bath of a salt of the same metal, and passes an electric current simultaneously through the whole series. He thereby dissolves the front or positive surface of each sheet as an anode, and simultaneously deposits pure metal upon its back or negative surface as a cathode, until nearly the whole of the crude metal has been dissolved, its impurities separated either in a soluble or insoluble state, and an equivalent weight of the pure metal deposited, the residuary layer of crude metal is then stripped off (*Jour.* Soc. Chem.

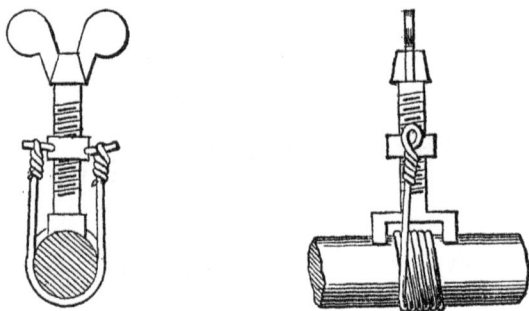

FIG. 65.—Connections of Electrodes.

Industry, Vol. VII., p. 390, 1888). This process is used by the "Bridgeport Copper Company," Bridgeport, Connecticut, U.S.A. (*see* pp. 175—176).

The Anodes.—In all ordinary cases these are formed of crude copper, usually "Chili bars," containing about 96 per cent. of that metal, and small amounts of sulphur, arsenic, antimony, bismuth, gold, silver, tin, iron, &c. (*see* pp. 188—189). They are sometimes cast of the annexed form (Fig. 66) with two hook-shaped projections, by means of which they are supported, one of the hooks resting upon the positive conducting bar of copper on one edge of the vat, and the other upon an insulating strip of dry wood, previously soaked in paraffin, on the opposite edge. Unlike the cathodes, they do not always require conducting bars across the vats to support them, and they are not

wholly immersed.   At Biache they are suspended by means of
two copper hooks, and two holes in the plates, to the metal
cross-bars of the positive frame, and are coated at their upper
parts with varnish.   At Pembrey they are supported by ribbon
hooks. At Casarza, being heavy and fragile, they are in addition
supported by two wooden ribs fixed upon the bottom of
the vat.

The usual thickness of each anode is about one-half or three-
fourths of an inch, but varies in different works ; at Casarza it
is 3 centimetres ($= 1·2$ inches), because each anode is composed
of mineral sulphides ; at Marseilles, about 1 centimetre.   It is
thicker at the upper part than at the lower, because the corro-

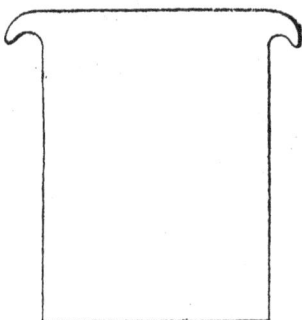

Fig. 66.—Anode.

sion there is more rapid. In order to prevent the anodes corroding through at the line of surface of the liquid and falling to
the bottom, they are sometimes coated at that part with varnish
containing some yellow chromate of lead.   The weight of each
anode varies greatly in different works, because of the considerable difference of dimensions.   At Marseilles it is " 11·4 kilogrammes" ; at Pembrey, about " 27·5lbs." ; at Casarza, 176lbs.
Large anodes are very inconvenient, on account of their great
weight, which renders it difficult to clean their points of contact with the conductors that support them ; they are also
awkward to handle, and are more liable to break when partly
corroded away.

At Pembrey each anode was "·625 metre long, ·175 metre wide, 1·25 centimetres thick, and immersed ·5 metre in the liquid." It exposed "·178 square metre" of active service on its two sides to the current, and the total amount of such surface in each vat was "2·8 square metres" (= 30 square feet); and subsequently five rows of narrow anodes in each vat, exposing a total of about 24 square feet of active surface per vat. At Hamburg the amount of surface of anode in each vat of the 120-series (*see* p. 160) was "15 square metres" (= 161 square feet), and in each one of the 40-series "30 square metres" (= 322·2 square feet), or in the whole forty vats "1,200 square metres." At Marseilles each anode was "·68 metre long, ·15 metre wide, and was immersed 58 centimetres in the liquid," and the amount of anode surface in each vat "22·5 square metres" (= 241·6 square feet), or in the series of forty vats "900 square metres." At Biache there were 88 anodes in each vat, in 22 rows of 4, each anode being "70 centimetres long, 15 centimetres wide, and averaged 1·0 centimetre in thickness," and the total amount of active anode surface, including both sides, in each vat was "18·5 square metres" (= 198·7 square feet). The amount of anode surface per vat in different works therefore varies from 24 to 322 square feet.

In the case of anodes formed by casting the crude matte of sulphides of iron and copper, which is very fragile, the ends of two strips of soft sheet copper are inserted in the liquid during fusion, and become firmly attached when the substance solidifies; these strips are used as suspenders, but the weight of the anodes is chiefly supported by two wooden ribs fixed upon the bottom of the vat. The anodes of matte at the Casarza works are "30in. long, 30in. wide, and 1¼in. thick," and require to be cooled very slowly after being cast, otherwise they fall to pieces. The casting of such anodes is difficult and uncertain; they are also apt to crumble and fall to pieces in the solution before they are all dissolved, and this causes great trouble.

Composition of Crude Copper.—The following analyses are copied from "Percy's Metallurgy," 1861, pp. 325, 361, 362, 422 :—

Black Copper is composed of—

Copper	95·45	89·13	92·83
Iron	3·50	4·23	1·38
Lead	?	·97	2·79
Silver	·49	?	·26
Zinc, nickel, cobalt	?	3·98	1·05
Sulphur	·56	1·07	1·07
	100·00	99·38	99·38

Pimple Copper—

Copper	89·4 to 95·6
Iron	·3 ,, 2·4
Sulphur	·4 ,, 2·5
	90·1 100·5

Blister Copper—

Copper	97·5	98·0	98·5	Copper 98·4
Iron	·7	·5	·8	Iron ·7
Tin and antimony	1·0	·7	·0	Nickel, cobalt, manganese ·3
Sulphur	·2	·3	·1	Tin, arsenic ·4
Oxygen and loss	·6	·5	·6	Sulphur ·2
	100·0	100·0	100·0	100·0

The copper usually most suitable for electrolytic refining is that free from antimony, bismuth, tin, and arsenic, containing a considerable proportion of silver, say 100 to 200 ounces per ton, and more or less gold. At the Hamburg works, bullion containing a large proportion of gold is refined. The entire product of the "Parrot Silver and Copper Company" of Butte, Montana, is electrolytically refined by the "Bridgeport Copper Company," Connecticut, and contains about 99 per cent. of copper, and from 70 to 100 ounces of silver per 2,000lbs. of copper, the whole of the precious metals being recovered.

The Cathodes.—In all cases they are made originally of very pure rolled sheet copper, and subsequently by electro-deposition. Usually the two outermost ones in a sufficient number of the vats are reserved to form new ones upon, the deposits being stripped off when they have acquired a suitable thickness; in this way a number may be made during the formation of a single thick intermediate plate.

When they are formed by this method, the inner surface of the copper is either slightly oiled or sprinkled all over with good conducting plumbago, and the surface thoroughly brushed in order to prevent adhesion of the deposited metal; the edges are then dipped into melted pitch, and the back surface well varnished and smeared all over with tallow, and the plate suspended as an outermost cathode in a vat. The rate of deposition must be moderate, otherwise the metal formed will not be tough. When the coating has acquired a thickness of about 1 millimetre, or $\frac{1}{25}$th of an inch, which occupies from two to fourteen days in different works, it is stripped off to form a cathode. The cathode thus formed, after having been used to deposit a thick plate upon, remains in the body of the plate, and is subsequently melted with it. The original sheets of copper upon which these cathodes are formed, after each formation of a cathode, require to be oiled or black-leaded afresh, and after several times using they must be re-varnished. In one refinery a small 5-unit dynamo is wholly employed in making cathodes; it deposits about 13lbs. of copper per 24 hours in each vat of a single series of 50. The weight of such a thin deposited cathode is usually about $2\frac{3}{4}$ to 3lbs.

When the intermediate or ordinary cathodes have acquired a total thickness of about $\frac{1}{2}$in., i.e., $\frac{1}{4}$in. on each side, which occupies from three or four (and in some refineries less) to twenty weeks, according to the rate of deposition, they are taken out to be melted and converted into wire, &c. In English refineries they are removed in about three weeks.

The usual method of connecting the sheet cathodes ready for suspension in the vats is by means of two clips of strip sheet copper of the annexed form (Figs. 67, 68, 69), and about one inch wide. They are, in nearly all refineries, wholly immersed in the liquid, and are supported by copper tubes or bars placed across the vats, one end of each tube resting upon the negative main conductor on one edge of the vat, and the other upon an insulating strip of wood or gutta-percha on the other. At Biache the thin cathodes are simply folded at the top, so as to hang upon the cross-bars of the negative frame.

Each cathode in the works at Casarza is placed in a wooden frame to keep it from touching the anodes, and is suspended by two copper bands, one of which is attached to the negative conductor ; when the deposit is sufficiently thick to remain flat, the wooden uprights are taken away, and when the cathodes are about " one-fifth of an inch " thick they are removed to be melted, &c.

Large cathodes are not very convenient, because they cannot be easily kept flat and equidistant from the anode at all parts, nor so readily stripped of their coatings when they are used to form new ones. At Pembrey each cathode was " ·425 metre long, ·4 metre wide, ·8 millimetre thick, and weighed 1·3 kilogrammes ; " and subsequently six in a row, each row exposing about 3 square feet of surface on each side, and weigh when taken

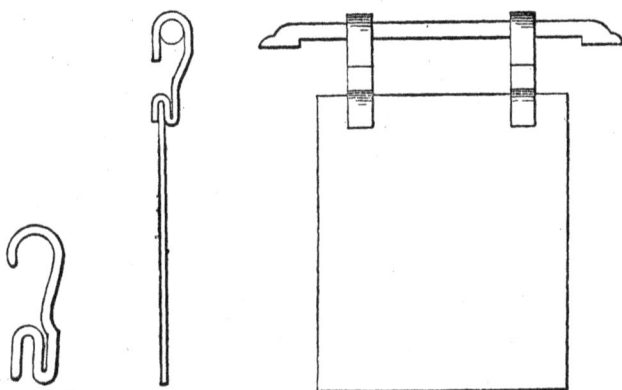

FIG. 67.   FIG. 68.   FIG. 69.
Mode of Suspending Cathodes.

out about 75lbs. At Marseilles they were ·68 metre long, ·15 metre wide, ·5 millimetre thick, and immersed 58 centimetres in the liquid. At Hamburg they were about 1 millimetre thick, and the total amount of cathode surface in the 240 vats was " 3,600 square metres." At Biache the cathodes were 85 centimetres long, 18 centimetres wide, and 1 millimetre uniform thickness. At Casarza they were " 27½in. long, 27½in. wide, and ·1in. thick."

Amount of Cathode Surface.—The total amount of active cathode surface in each vat is usually about the same as that of active anode surface. From the best available data, I have

calculated the amount in each vat of a series at different works
to be as follows :—Pembrey, with 48 vats in series, 32·9 square
feet (with the 100 or the 200 vats the amount of surface in
each is about 24 square feet); at Hamburg, with the series of
120 vats, 161·7 ; at Casarza, with 12 vats, 163·33 (*see* p. 158);
Biache, with 20 vats, 217·28 ; Marseilles, with 40 vats, 242·55 ;
Hamburg, with 40 vats, 323·4.   From the data, that at Oker
with each " C^{18} " dynamo, working 40 vats, "it requires five
months to deposit a piece of copper one centimetre thick"
(*Engineering*, Sept. 25, 1885, p. 306), and the daily deposit is
"771lbs.;" the total amount of cathode surface is about 5,904
square feet, and that in each vat is 147·67 square feet; and as
the total surface in the 12 vats worked by the " C^1 " dynamo
is "the same as in the 40," the amount of cathode surface in
each is about 492 square feet.   The amount of active cathode
surface, therefore, in each vat varies in different refineries
from about 24 to 490 square feet.

   The Main Conductors.—These are laid by the nearest route
from the dynamo to the vats, the positive one being fixed upon
one edge of the row of vats, and the negative one upon the
other edge throughout the series; and both are insulated in the
most perfect manner possible from each other, and from the
vats, all along their entire length.   In the works at Biache, each
vat being a foot or more distant from the next one in series, the
main conductors from vat to vat are arched to admit of passage
between the vats.   At Pembrey the conductors are laid across
the vats, the vats being end to end and near each other; the
positive conductor is laid upon the edge of one end of the
vat, and the negative one upon the edge of the other end.   In
some works they are laid upon the side edges, along the vats.
   The magnitude of cross-sectional area of the main conductors
should be consistent with the strength of the current and the
distance of the vats from the dynamo.   The larger the current
the greater should be the sectional area of the conductors, and
the lower its electromotive force the shorter their length.
They are formed of the best conducting copper, of sufficient
cross-sectional area not to become much heated by the current.
In some works they are of a semi-circular section (*see* Fig. 70),
in others circular, and in some rectangular.   They should have

a sectional area of about $1\frac{1}{2}$ or 2 square millimetres per ampere, or 1 square inch to about 320 amperes.

Those of Marchese's experimental works at Stolberg (*see* p. 235) were 10 millimetres diameter, and used to convey a current of 89 amperes; this is equal to ·882 square milli- metre per ampere, or 1·127 ampere per square millimetre. Those at Casarza are formed of cylindrical bars of copper, 1in. diameter, and convey a current of about 240 amperes. This equals 2·11 square millimetres per ampere, or ·474 ampere per square millimetre. At Pembrey, with a current of 320 amperes and 55 or 110 volts, they are solid circular rods 1·25in. diameter; this equals 2·474 square millimetres per ampere, or ·4042 ampere per square millimetre; they do not become sen- sibly heated.　At Oker, with the "C^1" type of Siemens's dynamo, yielding 1,000 amperes at 3·5 volts, working 12 vats in series,they have a sectional area of "25 square centimetres;"

FIG. 70.—Section of Main Conductors.

this equals 2·5 square millimetres per ampere, or ·4 ampere per square millimetre; and although the dynamos are placed close to the vats, the conductors become "sensibly warm;" but with the "C^{18}" type of dynamo, giving either 120 amperes at 30 volts, or 240 at 15 volts, working 40 baths in single series, the conductors are 12·0 millimetres thick, and the vats are at a distance of 40 to 50 metres from the dynamos. At Milton, with 5,000 amperes they are 6 square inches in section; this equals ·775 square millimetre per ampere.

The cheapest conductor is that one in which the money value of the loss caused by conversion of electric energy into heat, and the loss of interest upon the cost of the conductor, each during the same period of time, are equal.　The conductors should not be covered with felt or other substance to retain the heat, because that still further increases the loss by increasing the conduction resistance (*see* p. 31).

Preparation of the Depositing Solution.—The solution must
be strong, but not saturated. It is made by dissolving 24 ounces
of powdered blue vitriol in each gallon of hot "steam water,"
and when the salt is dissolved and the solution *cold*, adding gradually to it, with stirring, 10 or 15 ounces by measure of oil of
vitriol. The specific gravity of the mixture at 60°F. is usually
about 1·125 or 1·193, equal to 16deg. to 18deg. of Baume's
or 25 to 39 of Twaddell's hydrometer; at Pembrey it is 1·125,
at Marseilles 1·125 to 1·143, and at Biache 1·152, and kept
uniform. If it contains much more than 15 ounces of oil of
vitriol per gallon, oxygen is more liable to be evolved at the
anode during electrolysis, and this causes voltaic polarisation,
opposing the working current and wasting the electric energy.
The solution used at the North Dutch Refinery is said to be
a secret one, and to contain nitrates, but this is doubtful;
the nitrate solution is probably used in refining bullion (*see*
p. 242). During cold weather the solution may be slightly
warmed, not heated; this is best effected by means of a steam
injector, because this replaces water lost by evaporation. In
some works the temperature of the solution is kept at 20°C.,
in others at 25°C.; at Pembrey it is 23 to 24°C., at Marseilles
25°C., and maintained uniform. The temperature is recorded
daily. In hot weather it sometimes reaches 32°C. It is increased by the chemical action of the free sulphuric acid present, and by the occasional addition of that acid to compensate
that which has united chemically with the electrodes and with
the mud. The solution should be very pure in order to allow
rapid deposition of pure metal.

Circulating the Solution in the Vats.—Continual motion of
the liquid is very important (*see* p. 83); it largely increases
the range within which the strength of the current may be
varied without producing impure copper or causing polarisation.
Even with a pure solution which is not stirred, if the current
happens to increase there is a risk of depositing brittle copper;
the ways in which this occur have already been explained.
Warmth helps to keep the liquid mixed, because it reduces
its viscosity, increases its diffusive power, and causes liquid
currents. When the solution is too cold, steam is blown
through it before it enters the vats.

If the ordinary coppering solution is not constantly kept in motion it becomes stronger in copper at the bottom part and weaker and more acid at the top.   By taking a clear portion of it in a tall colourless glass vessel, immersing in it two narrow vertical copper electrodes, passing a strong current for some time without disturbance, and examining the solution with the aid of a suitable light, a stream of heavy liquid may be seen descending at the surface of the anode and one of lighter and less coloured solution ascending at the cathode.

In consequence of this inequality of composition of the liquid, the anodes dissolve away rapidly at the top without corroding as fast at the bottom, and sometimes dissolve through and fall to the bottom of the vessel, and the cathodes become rapidly thicker at the bottom without increasing in thickness at the top; and in extreme cases the anodes actually thicken by deposition of copper at their lower ends, whilst the cathodes are actually corroded and become thinner at their upper ones. An explanation of this has already been given (*see* p. 85).

The most serious consequences, however, are those of polarisation and the deposition of impure copper.   The former is attended by a direct and great waste of power in evolving oxygen at the anode, and the latter occurs by the deposition of some foreign metal, usually antimony or bismuth, at the cathode.

The usual method employed for mixing the liquid is to place each successive vat in a row upon a somewhat lower level, about one inch, than the following one, the solution in each vat being connected with that of the next by means of a vulcanised india-rubber syphon, reaching nearly to the bottom of the clear liquid in the lower one and regulated by a clamp screw, so that the solution may flow slowly through each row of a series by gravitation.   At one works, glass syphon tubes, bent slightly upwards at the upper or inlet end and bent downwards at the lower or exit end, are employed.   Another method, used at Casarza, is to have a gutter running along the edges of all the upper vats and a separate outflow from it into each of those vats, and a similar gutter along the edges of all the lowest vats to receive the overflows.   In order to decrease the loss of electric current passing from vat to vat in the overflow, it has been proposed to make the solution flow only at intervals, but this would not much diminish it; the loss also by

this cause is not very great (unless the pipes are short and of
large diameter), and does not occur at all in the terminal vats
in the first of these plans. At Casarza the liquid runs into each
vat by a right-angled lead pipe, the horizontal part of which,
lying upon the bottom of the vessel, has numerous small holes
in it for the exit of the liquid, and is protected by an open
wooden gutter from becoming covered with mud (*see* Figs. 71, 72).

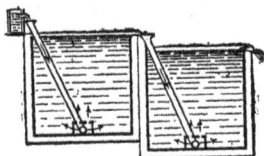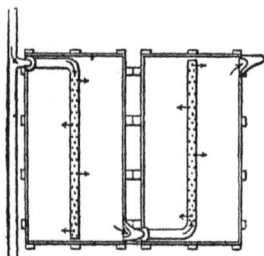

Fig. 71.   Fig. 72.
Mode of Circulating the Solution.

In each refinery the solution is supplied from an elevated tank,
into which the liquid is raised by means of a " Körting's injector " (Fig. 73).

At Biache " all the baths communicate together at the
bottom, so that their level is uniform," and when the liquid becomes too full of salt of iron, it is drawn off by means of leaden
syphons into a lower tank, and subsequently purified by crystalli-
zation ; it is then raised to the upper tank by means of a small
leaden injector and used again.

Inspection of the Vats, Electrodes, &c.—Constant examination of the vats, the electrodes, the solution, and the current is
indispensable, in order to detect and remedy leakages or short-
circuiting of current, to look for deposits of copper formed upon
the leaden lining of the vats, for solution spilled upon the vats,
for any heating of the conductors, &c.

The electrodes should be continually examined—1, for any
evolution of gas or deterioration of quality of the deposited
metal ; 2, for the degree of uniformity of corrosion of the
anodes and of deposition upon the cathodes ; 3, for anodes
which have become rotten, or have broken or fallen bodily off ;

4, for any fragments of anodes which have fallen between the
electrodes and short-circuited them, or have dropped or fallen
to the bottom and made contact with the leaden lining; 5, for
any nodules of copper projecting from the cathodes and touching the anodes; 6, for any electrodes which have gradually

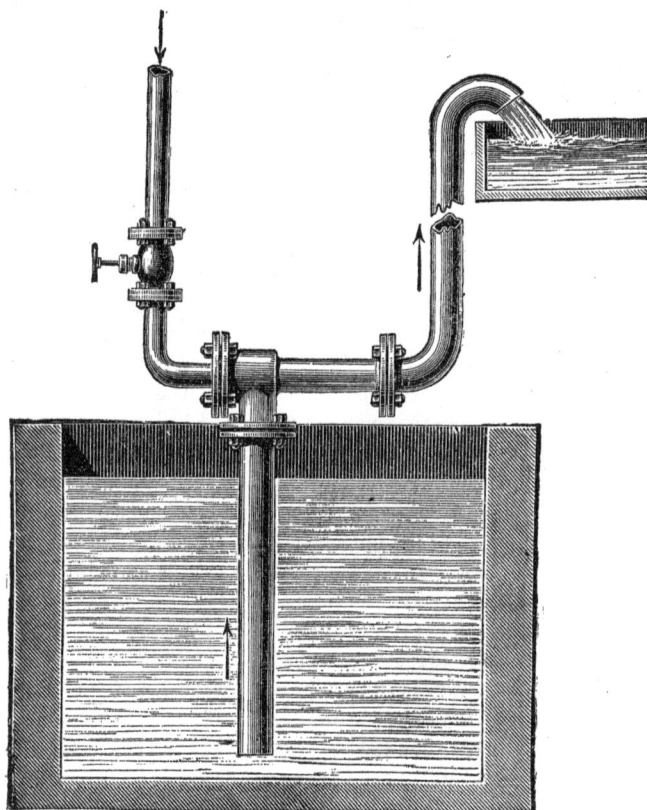

FIG. 73.—Körting's Injector.

bent, or anodes which have expanded and made short-circuits;
7, for anodes which have received a deposit of copper upon
their lower ends; and 8, for any defective contacts at the
points of connection and support: this is detected by their
becoming heated. Any gas visible at the anodes indicates

voltaic polarisation, counter-electromotive force, and waste of
electric energy ; and any at the cathodes means similar waste
of power and production of brittle copper, probably also deposition of impurities.

The anodes should be periodically taken out and cleaned,
and exchanged for new ones. How often depends upon the
quality of the copper in them and the rate of working, and
can be judged of by their appearance, and by any increase
of resistance in the circuit, caused by the coating of impurities upon them; this coating, being porous, does not however usually cause much resistance ; it consists chiefly of
sulphates, oxides, sulphur, carbon, &c. (see pp. 210, 213,
214). In some works they are changed for new ones in
about "three weeks," in others five weeks. Large ones are
usually removed by the help of an overhead travelling crane;
in removing them great care must be taken not to disturb
the mud or injure the lead lining of the vessels, that the
solution be allowed to perfectly drain from them before
removal, and not to drip upon the edges of the vats nor upon
the connections, also that none of the dirt from them gets upon
the cathodes, because it not only renders the subsequently
deposited copper impure, but also makes it rough and patchy.
Any cloudiness or muddiness in the solution should be noted.

The vats are cleaned out when it is desired to recover the
sediment or purify the solution, at no regular interval—say
once in six months.

Expenditure of Mechanical Power.—As by having the vats
and electrodes sufficiently large and numerous, a comparatively
limited amount of energy or of electric current may deposit an
almost unlimited quantity of metal (see pp. 151, 152), the
amount of copper deposited per horse-power, per 100 watts, or
per 100 amperes, is not a definite measure of economical
effect in all cases. We know that if there were no resistance, no polarisation or counter-electromotive force, and no
losses of current, 746 amperes at a difference of potential of
1 volt (and which equal 746 watts of electric energy, or
1 horse-power) would deposit 1·93 pounds of copper per hour
in each vat of an unlimited number in single series (see
pp. 126—128). But the commercial condition of greatest

economy of working limits the proportion of deposit per horse-power (*see* p. 152).

It has been correctly stated that "20 indicated horse-power will deposit three tons of copper in 144 hours," with a consumption of "less than three tons of coal," and that "the total cost of fuel for depositing three tons of copper by one large dynamo-electric machine amounts to only about 15 per cent." But this would not be an economical arrangement where motive-power or fuel is cheap, because the increased loss of interest upon extra cost of the vats and their contents would be much greater than the saving of cost of fuel. At one works, where fuel is cheap, 196 indicated horse-power is consumed in depositing 35½ tons of copper per week, but at Hamburg, where it is dear, the same amount of power deposited about 92½ tons. We must not forget that where fuel is cheap it is less expensive to consume more of it than to employ a greater stock of vats and of copper.

At the experimental installation at Stolberg "it is certain that less than 1 horse-power was consumed" in depositing "13·322 kilogrammes of copper each 24 hours ;" this equals 1·22lbs. per horse-power per hour. At Casarza, "two tons of copper were deposited per day," the entire motive power being derived from "two turbines of 50 horse-power each, and one of 75 horse-power ;" this equals 1·49 lbs. per horse-power per hour. According to Dingler's *Polytechnisches Journal,* Vol. CCLV., p. 201, the "daily output of copper" at those works "is 20 kilogrammes per horse-power" (*see* pp. 232—234) ; this equals 1·83lbs. per horse-power per hour. At Biache with a "Gramme machine" identical with "the larger one at the North Dutch refinery," consuming "16 horse-power," the total amount of copper deposited was "400 kilogrammes daily" ; this equals 2·29lbs. per horse-power per hour. At Pembrey, 196 indicated horse-power deposits about 509·75lbs. of copper per hour = 2·6lbs. per horse-power per hour. At Balbach's Works, New Jersey, a "25 horse-power" engine deposits "1 ton per day," equalling 3·73lbs. per horse-power per hour. At Hamburg, with the large Gramme machine, consuming "16 horse-power," the total quantity of copper deposited was "30·5 kilogrammes per hour" ; this equals 4·18lbs. per horse-power per hour. At Marseilles, with a "No. 1" Gramme dynamo, consuming "5 horse-

power," the total amount of copper deposited was " 10·4 kilogrammes per hour "; this equals 4·576lbs. per horse-power per
hour.   At the Hüttenwerke, at Oker, with a " C^1 " Siemens'
dynamo, consuming " 5 horse-power," the total quantity of copper deposited was " 300 to 350 kilogrammes (= 771lbs.) per 24
hours "; this equals 5·48 to 6·4lbs. per horse-power per hour.
And at Hamburg, with two " No. 1 " Gramme machines, connected in series, consuming " 12 horse-power," the total amount
of copper deposited was "900 kilogrammes per 24 hours"; this
equals 6·87lbs. per horse-power per hour.   At Bridgeport "each
60 horse-power dynamo deposits 11,000lbs. per 24 hours;"
this equals 7·63lbs. per horse-power per hour.   According to
these data 1 horse-power is expended to deposit from 1·22 to
7·63lbs. of copper per hour in different electrolytic refineries.
These considerable differences of result are largely due to differences of resistance in the vats, and of number and size of vats
in series; and the larger proportion of copper deposited per
horse-power at Hamburg than at the other places is obtained at
the expense of greater cost of interest upon a larger investment
of capital in vats, copper, and electrolyte, and not necessarily
at a less total cost per given weight of copper deposited.

Expenditure of Electric Energy.—The proportion of copper
deposited to electric energy consumed varies in every different
works.   At Bridgeport, Connecticut, " each Mather dynamo
yielding 40,000 watts deposits 11,000lbs. per day," equal to 1lb.
per hour to 87·28 watts.   At Hamburg, with the two " No. 1 "
Gramme dynamos connected in series, and yielding 300 amperes
at 27 volts, equal to 8,100 watts, the total quantity of copper
deposited was " 900 kilogrammes in 24 hours;" this equals 1lb.
per hour to 98·18 watts.   At the Hüttenwerke, Oker, with the
" C^1 " dynamo, giving " 1,000 amperes at 3·5 volts," equal to
3,500 watts, the total amount of copper deposited was "about
350 kilogrammes a day "; this equals 1lb. per hour to 109·4
watts.   At Hamburg, with the large Gramme dynamo giving
12,000 watts, the quantity deposited was " 30·5 kilogrammes
per hour"; this equals 1lb. per hour to 178·8 watts.   At Biache,
with a " Gramme machine identical with " the last mentioned,
the amount deposited was " 400 kilogrammes daily "; this
equals 1lb. per hour to 327 watts.   At Pembrey, with a long

armature dynamo yielding 38,500 watts, 4,000lbs. of copper are deposited daily = 1lb. per hour to 231 watts. The number of watts of electric energy expended in depositing 1lb. of copper per hour, therefore, varied from 87·28 to 231 in these different works.

Sources of Loss of Energy.—There are many sources of loss between the prime mover and the final deposition of copper, and these have to be diminished as much as possible. In ordinary electro-refining of copper the chief sources of loss of electric energy are :—1st, in the dynamo by conversion of it into heat ; 2nd, in the conductors by a similar cause ; and 3rd, in the vats by various circumstances ; the latter is usually the largest. There is very little leakage, not more than 1 or 2 per cent. in vats lined with pitch.

The amount of waste in the dynamo is roughly indicated by the degree of heating of that apparatus. This heating is partly produced by conduction resistance in the wires, and partly by induction currents (and resistance to them) in the conductors and in the metal portions of the magnets and armatures. Some, at least 1 or 2 per cent., of the electrical energy is necessarily expended in exciting the field-magnets, and not less than 2 or 3 per cent. in the armature. With the most efficient dynamo, out of 100 units of mechanical energy imparted to it, between 93 and 94 are available as electric energy in the external circuit.

After working a long time, the energy of the current from the dynamo and the difference of potential at its terminals may largely diminish in consequence of short-circuiting, through metallic dust worn off the brushes and commutator getting between the sectors of the latter, or through oily matter between those divisions becoming charred by the sparks ; also the segments of the commutator may not continue perfectly insulated from the axle, or the terminals of the dynamo from its bed-plate, or the armature conductors from the iron beneath them, etc.

As no substance is a perfect conductor even to the feeblest current, total avoidance of heat of conduction resistance in any portion of the circuit is impossible. In the main conductors it should be reduced by having them as short and thick as will prevent any undue heating (*see* p. 134); the vats, therefore, should be as near as convenient to the dynamo, and the surfaces of contact of the conductors should be large and perfectly clean.

In the vats the chief sources of loss are—1, by voltaic polarisation and counter-electromotive force; 2, by evolution of gas at the electrodes; 3, by chemical corrosion of the deposited metal (*see* pp. 65-66, 120-124); 4, by conduction-resistance of the liquid; 5, by energy necessarily consumed in conveying metal from the anode to the cathode, and of oxygen in the opposite direction; 6, that expended in causing visible movements in the electrolyte; and 7, that consumed in deoxidising persalts of iron; the latter, with crude matte, is considerable. The polarisation of the electrodes has been estimated to "consume mechanical energy varying between 5 and 10 per cent. of the total work." Several per cent. of the current which enters each cathode practically yields no deposit, in consequence of chemical corrosion of the deposited metal (*see* pp. 121—125).

In addition to these sources of loss, there are leakages of current by imperfect insulation of the vats and conductors from the ground, by portions of current getting from one conductor to another through the damp wood of the vats, through films of copper solution spilled upon the vats, from one vat to another through the connecting tubes, and through the damp woodwork of the supporting stage beneath the vats. The copper solution is occasionally splashed or dropped upon the edges of the wood and soaks into it, or it creeps over the edges and upon the conductors by capillary action. Constant examination is therefore necessary in order to detect leakages and short-circuits; the locality of the former is often indicated by a rise of temperature of the conductors, perceptible to the touch at particular places, also by reduction of the damp film of blue salt to red metallic copper.

There are also losses of energy by unnecessary resistances producing heat at defective contacts, besides occasional great waste of power by accidental short-circuiting. The amount of energy therefore absorbed is far from being nothing, and each vat is usually considered to absorb one-third of a volt, and requires provision of a total of about half a volt to overcome all resistance and polarisation.

Waste by leakage of current and by short-circuiting must be especially attended to. Their existence is indicated in a general way by a diminution of the total resistance in the circuit, and is accompanied by an increase in the total strength of current and a fall of potential, as shown by the ammeter, voltmeter, or other indicator.

The waste by conduction resistance in the vats and connections should be reduced in every possible way, by using sufficiently large electrodes, by reducing the distance between them to a minimum, by maintaining the maximum safe degree of acidity of the solution, and by keeping the liquid sufficiently warm. Defective contacts should be carefully avoided by frequently cleaning them, and by keeping the air of the depositing room pure and free from acid vapour. Anodes imperfectly connected will not dissolve, and cathodes badly attached will not receive a full deposit.

Resistance of the Solution.—As a certain proportion of the energy of the current is expended in overcoming the resistance of the liquid, it is desirable to know the magnitude of this obstacle and to diminish it as far as possible. In order to ascertain the amount of this resistance we require to know the composition and temperature of the liquid, and the sizes and distances asunder of the electrodes. The larger the electrodes, and the nearer they are together, the less the resistance. The resistances in ohms of a cubic centimetre, both of a solution of blue vitriol and of dilute sulphuric acid of various percentages and specific gravities, and at different temperatures, are given in the Tables on pages 30—35. As the resistance of moderately-dilute sulphuric acid is much less than that of a solution of cupric sulphate, and mixtures usually conduct better than the mean of their ingredients, the resistance of the blue vitriol solution is considerably diminished by the addition of the free acid, and that of a cubic centimetre of the mixture at 20°C. is about 20 ohms. The resistance, therefore, of two electrodes 1 metre wide and 1 metre long ($=10,000$ square centimetres) at 1 centimetre distance asunder would be $\dfrac{20}{10,000} = ·002$ ohm, and at 5 centimetres asunder $= ·01$ ohm; as, however, the current is not strictly confined to a section of the liquid equal to the actual

area of the electrodes, but in its passage between them spreads
out on all sides laterally (*see* Fig. 27, p. 80), the resistance is
thereby somewhat lessened.   The resistance of the baths must
not be much further reduced by larger addition of acid, because
that tends to the more ready production of polarisation and of
gases at the electrodes.   The resistance of the ordinary copper
refining solution between electrodes 1 metre square, at 1 centimetre apart, usually varies from " ·01 to ·03 ohm." At Marseilles, with 22 square metres of each electrode in each bath, and
the electrodes 5 centimetres apart, the resistance = " ·00046
ohm," and of the 40 baths ·01840 ohm.   At the North Dutch
Refinery the total resistance of 240 vats arranged in parallel as
120 was " ·1008 ohm, or ·00084 ohm per vat."   At Oker the
" total resistance " of the " ten or twelve large vats, and of the
conductors is about ·0055 ohm."   At Pembrey, with the plates
about 3in. apart, a density of current of about 10 amperes per
square foot, and 100 vats in series, the total resistance is about
·16 ohm, and with the 200 vats ·32 ohm, or about one six-hun-
dredth of an ohm per vat.   Under favourable conditions the resistance is about ·00140 ohm per vat.   When the electrodes are
about 2in. apart and the density of the current is about 8 amperes
per square foot, each vat usually requires an electromotive force
of ·33 volt to overcome the resistance and polarisation at 20°C.

Examination of the Current.—The current should be frequently examined and kept uniform.   An excess is injurious
both to the dynamo and to the quality of the deposited copper,
it may also indicate short-circuiting, whilst a deficiency may be
caused either by defective contacts or by badly conducting
coatings upon the anodes.   If, however, a decrease of resistance
by short-circuiting happens at the same time as an increase by
other causes, the total main current may be unaffected.

Detection of Resistances and of Leakages.—In every electrolytic refinery frequent measurements are necessary, both of the
potential difference of the baths and of the strength of current;
the former serve to detect resistances, and the latter show
the rate of deposition.   In addition to the detection of local
resistances by a manifest rise of temperature at particular
points of the circuit, and of excess of current in the vats by

visible evolution of gas or darkness of colour of the deposited
copper, a voltmeter or other instrument is employed to detect
changes in the degree of potential in the main circuit, caused
either by polarisation or by increased or decreased total resistance in the external circuit, or in the dynamo.   A shunt galvanometer or an ammeter (Fig. 74*) is also used to measure the
amount of current passing at any point of the external circuit.

FIG. 74.—The Ayrton and Perry Ammeter.

In addition to these, the temperature of the liquid in the vats
is tested at different points by means of a thermometer.   A
switch or cut-out is also used in case of necessity (Figs. 75 and
76†), and in some refineries an automatic cut-out is employed.
A shunt and resistance is often worked by hand.

* This instrument is manufactured by Messrs. Latimer Clark, Muirhead
and Co., of Regency-street, Westminster, London, S.W.

† These switches are manufactured by Messrs. King, Mendham and Co.,
of Narrow Wine-street, Bristol.

Amongst the various instruments invented for measuring the difference of potential, strength of current, and resistance are Siemens's torsion galvanometer and electro-dynamometer (see Fig. 10, p. 22), Ayrton and Perry's voltmeters, ammeters and ohmmeters, Crompton's indicator, Cardew's voltmeter, Cunyngham's ammeter and voltmeter, &c. ; and of those invented for stopping or shunting the current, are Weston's electric governor (which, when the main current is too strong, shunts it through a resistance), Goolden and Trotter's magnetic switch and cut-out, the switches of Siemens, Sawyer, and others. The torsion galvanometer of Siemens and Halske is an instrument generally useful ; it can be employed for measuring the electromotive force of the dynamo,

FIG. 75.—Lever Switch.   FIG. 76.—Plug Switch.

the counter-electromotive force of any bath, or, by measuring the potential difference at the extremities of a known resistance through which the current is flowing, the strength of the latter is obtained. In some cases a large form of Siemens's electro-dynamometer is used, capable of being inserted in the main circuit and of measuring a current up to 1,000 amperes. The advantage of Ayrton and Perry's spring ammeters and voltmeters is, that they are direct-reading without calculation, and without having to multiply by a constant number, or to refer to a table of values. They have also no steel magnet, the power of which is liable to change.

Measurements of resistance, determinations of difference of potential at the terminals of the dynamo, ascertaining strength of current by measurements of difference of potential between fixed points upon the main conductor whose resistance is known, and difference of potential between the electrodes in each vat, are usually sufficient for practical purposes.

Fig. 77 shows an arrangement sometimes used in small depositing works for regulating and testing the main current, independent of the separate regulating and testing the potential and strength of current of each vat or row of vats. A is an ampere-meter, V a volt-meter, R a resistance of iron hoop or wire to act as a current regulator; K is a key to throw the voltmeter into or out of action, and S is a switch to short-circuit all the vats. Resistance-regulators are sometimes formed of ribbons of German-silver wire-gauze or of thin sheet-iron.

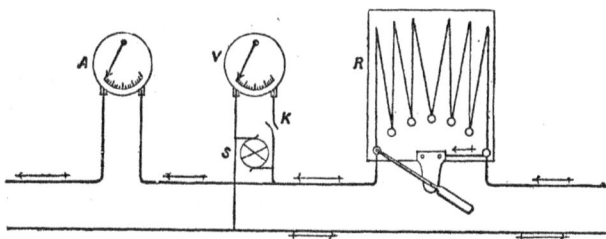

FIG. 77.—Arrangement for Regulating and Testing Main Current.

Limit of Rate of Deposition.—On account of the great value of the stock of copper and solution under treatment, it is a chief object of the refiner to deposit the metal as fast as possible upon the smallest amount of cathode surface, but the maximum speed is limited by the kind of impurities in the liquid and their amount, and varies with each batch of unrefined copper. If bismuth, arsenic, and antimony are not present in a soluble state, the speed of deposition may be greater. The unrefined metal ought to exceed 98 per cent. in purity, and be quite free from bismuth and antimony, in order to deposit at a rate of 10 amperes per square foot. The practical density of current is also limited by waste of energy due to polarisation, which increases at a rate approaching that of the square root of the strength of current, but somewhat less in consequence of the rise of temperature of the liquid and its diminished resistance. A maximum of 10 amperes per square foot (or 9·84 ounces of copper deposited per square foot per 24 hours) is commonly allowed; but if short circuits occur in the vats, some of this does not actually go into all the cathode surfaces. If the rate of deposition is too rapid, there is great loss of value in the quality of copper deposited, and increased loss of energy in overcoming incipient polari-sation ; and if it is too small, there is the extra loss of interest upon capital invested in copper and electrolyte during the longer period. The economic rate of working, therefore, is that at which the money values of these two losses are equal. Polarisation commences before any visible evolution of gas.

Density of Current and Rate of Deposition.—This varies largely in different refineries, and is chiefly governed by the kind and amount of impurities dissolved in the liquid, and the required degree of purity of the deposited copper. It usually varies between 5 and 10 amperes actually passing into each square foot of active cathode surface (*see* below). If the solu-tion is not stirred, 8 efficient amperes per square foot should not be exceeded because of incipient polarisation, unless the liquid is quite free from easily reducible metals.   With a mixture of 3 parts of saturated solution of blue vitriol, and 1 part of dilute sulphuric acid, 1 to 10 of water, one writer, Mr. J. T. Sprague, in his book on "Electricity," 1875, pp. 284, 285, extends the limit to "1½ units in 10 hours, upon 1 squa·e inch of surface," the "unit" being "nearly 32 grains of copper;" "1·5 units" would, therefore, equal about 38 amperes per square foot. Another writer, Mr. A. Watt, in his "Electro-Metal-lurgy," first and second editions, 1886, 1887, p. 424, states that "the quality of copper is found to be uniformly pure when Mr. Sprague's limit is not exceeded."

From published and other statements of the quantities of copper deposited daily upon given amounts of cathode surface in different works, I have calculated the densities of current employed, and the rates of deposition, as follows :—At Ham-burg, with the series of 120 vats, ·823 ampere per square foot, and ·8098 ounce per square foot per 24 hours. At Marseilles, with 40 vats, ·913 ampere, and ·998 ounce. At Stolberg, with 58 vats, 1·482 amperes, and 1·458 ounces. At Casarza, with 12 vats, 1·84 amperes, and 1·81 ounces. At Biache, with 20 vats, 1·9096 amperes, and 1·879 ounces. At Hamburg, with the 40 vats, 2·008 amperes, and 1·976 ounces. At Oker, with the 40

vats, and "C^{18}" dynamo, 2·124 amperes, and 2·09 ounces ; and
with the 12 vats and "C^1" dynamo, the total amount of
cathode surface, of electric energy, of copper deposited daily,
and density of current, were the same as in the 40 vats.   At
Messrs. Balbach's Works, New Jersey, the density is "10 amperes per square foot."   And at Pembrey, where fuel and
motive power are relatively cheap, with the 100 or the 200 vats
in series, 10 amperes per square foot, and 9·84 ounces per
square foot per 24 hours.   The range of variation in these cases
is from ·82 to 10 amperes, and from ·8098 to 9·84 ounces per
square foot per 24 hours.   In some works the latter proportions
are exceeded.

Thickness of Deposit.—According to these data, the rate of
increase of thickness of deposit varies from ·001 millimetre
per hour or ·156 millimetre per week of 156 hours, with the
two "No. 1" Gramme machines at the North Dutch refinery,.
to ·0143 millimetre per hour, or 2·231 millimetres per week, at
Pembrey.   "38 amperes per square foot" (see p. 208) would be
equal to 15·33lbs. of copper in 156 hours on that surface, or an
increase of about ·33 inch, or 8·38 millimetres in thickness per
week.   Ten amperes per square foot on each side of a cathode
will produce a plate fully half an inch thick in about three
weeks (see Appendix).

Economy of Working.—The entire economy of working
depends essentially upon cheapness of motive-power and minimum investment of capital, and the information already given
shows that nearly every attempt to improve the process, in
the direction of such economy, involves a choice between two
evils.   For instance, if we try to save horse-power by enlarging
the electrodes and adding to the number and size of vats in
series, we rapidly increase the loss of interest upon capital
expended in stock of copper, solution, plant, working space, &c.
If we increase the density of current with the intention of
working more rapidly with the same stock of copper, solution,
&c., we increase the loss of energy by causing polarisation, and
run a risk of depositing impure copper.   And if we attempt
to diminish the resistance by placing the electrodes nearer together, or by quickly stirring the solution, we get dirt upon the

cathodes, or short-circuiting occurs; and if we endeavour to decrease it by considerably heating the liquid, we rapidly increase the chemical corrosion of the deposited metal, and quickly alter the chemical composition of the liquid; or if we diminish resistance by large additions of acid, we promote chemical corrosion, polarisation, and separation of gases.

Difference of potential required is a large element in the economical deposition of copper. Whilst the same number of amperes of current passing through the ordinary solution deposits the same amount of copper in all cases, the necessary degree of difference of potential to overcome resistance, &c., varies in different cases; and the greater this difference the larger the cost: double the potential requires double the motive-power. It is desirable, therefore, to lower the difference of potential as much as possible by diminishing the resistance throughout the circuit, and by avoiding all polarisation and counter-electromotive force.

The degree of electromotive force required varies with every different kind of solution, and consequently of anode. It varies according as the anode is composed of pure copper, "black copper," or "copper matte," and whether the anode is clean, or covered with an adhesive badly-conducting coating; it is also much greater when gas is evolved at either electrode.

Influence of Impure Anodes upon the Liquid.—I will only speak of some of those substances which are more or less likely to be present in "black copper," "pimple copper," "Chili bars," "blister copper" (see p. 189), containing from 89 to 98 per cent. of copper, and in cruder regulus of reduced pyrites of iron and copper. Those impurities include antimony, arsenic, bismuth, cadmium, carbon, cobalt, gold, iron, lead, manganese, platinum, silver, tin, zinc, suboxide of copper, sulphides of iron, copper, and silver in the particles of unreduced pyrites of those metals, alumina, lime, magnesia, silica, and alkalies in the enclosed portions of slag. Black copper has been found to contain "1·23 per cent. of arsenic, 1·0 of iron, ·54 of sulphur, ·4 silver, and ·011 of gold" (*Engineering*, 1885, p. 306).

In an acidulated solution of blue vitriol, the following of those substances, viz., cadmium, cobalt, iron, zinc, sulphide of iron. alumina, magnesia, and alkalies of the slag, dissolve readily; antimony, arsenic, bismuth, tin, and silica dissolve imperfectly, and partly fall to the bottom; carbon, gold, platinum, and sulphur from the pyrites are insoluble, and pre-cipitate entirely; lead is converted into sulphate, which almost wholly precipitates; silver is changed into chloride, provided a soluble chloride is present, which is usually the case, and is entirely thrown down; suboxide of copper, and the sulphides of copper and of silver, also precipitate. The lime is all changed into sulphate, a small quantity only of which enters into solution, whilst the remainder subsides.

One effect of using an impure metal or an alloy as an anode is, that its different constituents are corroded unequally, both by the ordinary chemical action of the liquid and by the influence of the current. By both these actions the most electro-positive metals are attacked first, and the others in succession, in accordance with the thermal law which governs such actions (see p. 49), and those which are not corroded at all, together with those which form insoluble compounds, either remain upon the surface as a coating, or fall to the bottom as mud. In consequence of this irregularity of action, the anodes are often corroded deeply in places, even whilst containing much metallic copper in their coatings, become rotten, and pieces of them fall to the bottom, and this occurs the most with very impure metal, and especially with cupreous matte. When the anodes are made by fusion of copper pyrites, they often swell on the sides towards the cathode after a few days of electrolytic action. If also the liquid is not kept homogeneous, impure ones are rapidly corroded, and sometimes cut through at the surface of the solution and fall to the bottom.

If arsenic is present, it partly dissolves as arsenious or arsenic acid, and gradually unites with some of the constituents of the mud to form insoluble compounds; arsenious acid is sparingly, and arsenic acid freely soluble. It also partly forms arseniates upon the anode, which are non-conductors, and by adhering to that electrode, offer resistance to the current.

Antimony partly dissolves, especially in acid solutions, and partly forms an insoluble basic salt, some of which adheres to the anode, oxidises and increases in weight, and some falls to the bottom; any oxy-salt of antimony appears as a white cloudiness when the liquid is sufficiently dilute.

Bismuth behaves much like tin and antimony; it partly forms an insoluble basic salt, which precipitates, and partly dissolves as an acid salt, which ultimately also becomes basic and falls.

Tin behaves much like antimony ; it forms a mixture of proto-salt and basic salt, some of which adheres to the anode and some dissolves ; the proto-salt is white when dry, and absorbs oxygen from the air rapidly to form persalt. The presence of dissolved tin in a neutral solution is said to greatly improve the quality of the separated copper, without the tin itself being deposited ; this, however, probably only occurs under particular conditions, because in some experiments I made for the purpose, this effect did not take place. The presence of tin in the anode increases the electromotive force of the current, because tin is more electro-positive than copper in dilute sulphuric acid (see p. 56).

Lead, being also more electro-positive than copper, is attacked before it, and aids the electromotive force ; but its sulphate being very insoluble, falls nearly wholly to the bottom, and only a trace is dissolved. There is but little risk of lead being deposited with copper, not only because its sulphate is so very insoluble, but also because lead is electro-positive to that metal. Lead, by being electrolytically corroded in place of copper, and uniting with acid to form an insoluble salt, diminishes the amount of copper and of acid in solution.

If the anode contains metallic iron and manganese, those metals are dissolved before the copper, and being more electro-positive (see p. 56) assist the current ; they dissolve as proto-salts, and subsequently become oxidised to some extent, but not completely, to persalts by contact with the air. If, however, the current is very much stronger than that usually employed in refining copper, these proto-salts are also oxidised to persalts at the anode ; and in any case their persalts are reduced to proto-salts at the cathode by the current and cause great waste of power (see pp. 93-94, 118). The only way to prevent waste of energy in deoxidising persalts at the cathode is to keep those salts out of the solution. Iron by dissolving in place of some of the copper at the anode, takes up some of the acid, forms green vitriol, and diminishes the solvent power of the water and the amount of copper in solution.

Any suboxide of copper in the anode, being a non-conductor, falls to the bottom, and is subsequently slowly oxidised and dissolved as cupric sulphate by the free acid. If the anode contains much cupreous sulphide, some of it is decomposed by the current, its metal dissolving whilst its sulphur separates and subsides. If its amount is small (as it usually is in "Black Copper"), being a very inferior conductor it is unaffected by the current, and falls to the bottom as fast as the metal around it is dissolved.

In an anode containing copper, copper oxides and sulphides, the current goes most by the path of least resistance, i.e., very largely by the metal, very little by the sulphides, and much less by the oxides. If, therefore, the anode contains very little sulphide or oxide, the whole of the current travels through the copper, and the oxide and sulphide fall to the bottom; but if it contains much sulphide, some of the current passes through it. At the same time, independently of the action of the current, the liquid is always acting chemically upon each of these substances, whether they form part of the anode or of the sediment.

It is evident that when the anode contains a large variety of impurities the chemical actions going on at its surface must be numerous and exceedingly complex, and modify each other, and make it very difficult to accurately describe them. One effect of those numerous impurities is to cause an almost infinite number of local electric currents all over the anode, attended by corrosion and solution of all the more positive substances without the aid of the external current, thus gradually neutralising the free acid, and saturating the water with metallic salts.

Influence of the Impurities upon the Current.—Provided the anodes are wholly metallic, their lower degree of electric conductivity due to impurities has very little effect upon the total resistance of the circuit, because the resistance of the solution is so very much greater (see p. 25). Consequently, also, it has but little effect upon the kind of dynamo required.

Impurities in the anode, however, have in some cases a great effect upon the electromotive force and upon the strength of current, either by giving rise to polarisation and counter electromotive force, or by the formation or accumulation of badly-conducting substances upon the anode. The greater the number of impurities also the greater the risk of waste of energy.

It is a fact that the electromotive force, and consequently also the strength of current, is increased when the anodes contain as impurities metals such as zinc, cadmium, manganese, or iron, which are more electro-positive than copper in the particular liquid, because they constitute with the copper cathode a voltaic couple sending a current in the same direction as the working one. But this assistance to the current, except with iron pyrites, is usually so very small that it is hardly observable, and the advantage of it is more than counterbalanced by disadvantages. The presence of iron in the anode aids the current to only a small extent, but that of zinc helps it more.

Usually when fresh anodes are put in, the resistance of the vat gradually rises to a maximum, but when the film of impurities is removed from them the resistance falls at once slightly ; it also usually increases with the thickness of the coating, but only slowly because the layer is porous and saturated with the solution. The coating acts chiefly by preventing ready access of unexhausted liquid, and thus promotes polarisation.

When the solution contains persalts of iron or manganese those salts are reduced to proto-salts at the cathode, and a portion of the current is continually wasted in reducing them instead of being wholly employed in separating copper (see pp. 93-94, 118), and the copper costs more to refine than when no iron or manganese is present. As long as there is sufficient persalt of iron in the liquid, no hydrogen is evolved at the cathode. With a copper-depositing solution containing much salt of iron, the difference of potential of plates of pure copper of a single vat has been found by experiment to be ·22 volt, with an anode of black copper ·25 volt, and with one of copper pyrites ·5 volt, and when the solution was deficient in free acid, it was ·35 volt with black copper, and ·75 volt with copper pyrites (Kiliani, *Engineering*, Vol. XL., 1885, p. 306).

Effect of Impurities of the Liquid upon Purity of the Deposit.—The electrolyte is always less impure than the anode, because various of the impurities of the latter are either loosened by corrosion of the surrounding metal, and separate and fall to the bottom unchanged; or they are separated from their compounds by chemical or electro-chemical decomposition and subside; or they are converted by chemical action into insoluble compounds, which precipitate (see pp. 210-213).

The more impure the anode the less pure usually is the electrolyte. In addition to the proper constituents of the liquid, viz., blue vitriol, sulphuric acid, and water, there may be present in a dissolved state the sulphates of iron, zinc, manganese, cadmium, cobalt, aluminium, magnesium, calcium, potassium, and sodium, basic sulphates of antimony, tellurium, bismuth, and tin, and arsenious or arsenic acid.

By far the largest in amount of these dissolved impurities is the sulphate of iron; but the most objectionable ones are the compounds of arsenic, antimony, and bismuth, because they are the most likely to yield their metals at the cathode, and thus affect the purity and quality of the deposited copper.

During the progress of deposition the electrolyte continually becomes more impregnated with impurities, especially with iron, and, in much less degrees, also with zinc, manganese, cobalt, nickel, tin, arsenic, antimony, bismuth, aluminium, calcium (within small limits), magnesium, potassium, and sodium, and gradually poorer in free acid, and, sooner or later, requires to be purified.

Influence of Impurities of the Liquid upon the Current.— The greater the number of impurities in the electrolyte, the greater the risk of waste of electric energy. All ordinary sulphate of copper solution used in the electrolytic refining of copper contains, after a time, a greater or less proportion of dissolved ferrous sulphate, and this substance gradually combines with the oxygen of the air and becomes persalt. In such a mixture there is liable to occur a considerable waste of current and deficiency of deposited copper— 1st, because much of the electric energy is wasted in deoxidising the persalt at the cathode (see pp. 93, 94), and 2nd, because a solution of persalt of iron corrodes and dissolves copper (see p. 121); the free acid therefore decreases, and the liquid acquires greater specific gravity. The presence of green vitriol, however, prevents polarisation at the anode, because it absorbs oxygen.

In a solution of green vitriol partly oxidised by the air, and being suitably electrolysed with platinum electrodes, a mixture of iron and hydrogen is deposited at the cathode, and persalt is formed at the anode ; and, owing to absorption of oxygen from the air, the proportion of persalt increases (*see* p. 119).

Chemical Examination of the Solution.—The solution should be kept uniformly at the same temperature and degree of concentration, and be tested occasionally in order to ascertain whether it is of proper composition, and corrections applied accordingly. The proportions of copper, water, free acid, and sulphate of iron should be maintained within suitable limits. When the liquid gets too impure, it should be removed, evaporated, and crystallised so as to recover the blue vitriol, which is usually obtained sufficiently pure to form fresh solution. " Steam-water " or rain-water is suitable to dissolve the crystals. As 100 parts of water at 18° C. dissolve 60 parts of green vitriol but only 42·7 of blue vitriol, the latter crystallises out first during evaporation.

A convenient way of ascertaining the proportion of free acid present is to take a known volume of the cold solution, add to it gradually, with constant stirring of the copper liquid, a solution of caustic potash or soda of known strength, until a very small amount of precipitate occurs which fails to re-dissolve by persistent stirring, and calculate the amount of sulphuric acid necessary to neutralise the quantity of alkali used.

To determine the amount of copper in the presence of the large proportion of iron, and the small quantities of other metals likely to be present, various methods of chemical analysis may be employed. The following is an outline of one of them :—Take a known bulk of the solution, pass a stream of washed sulphuretted hydrogen gas through it as long as a precipitate occurs, and until it smells strongly of the vapour. Let the precipitate subside perfectly in the covered vessel, then either decant or filter away the clear liquid containing all the iron, and wash the precipitate unintermittingly with distilled water mixed with sulphuretted hydrogen water. Re-dissolve the precipitate without delay in aqua regia ; separate the liberated sulphur, burn it to recover a trace of copper, which must be re-dissolved, and precipitate the copper from the cold diluted liquid, either as metal by the electrolytic process, or as black oxide from the boiling hot liquid, by adding an excess of solution of caustic potash or soda with constant stirring; wash, dry, and heat to redness the black oxide of copper, 1 part by weight of which equals ·7985 part of copper, or 3·137 parts of crystallised blue vitriol.

To ascertain approximately the amount of iron, precipitate the filtrate from the sulphide of copper by an excess of sulphide of ammonium, wash the precipitate in a covered beaker with water containing a little sulphide of ammonium, re-dissolve it in hydrochloric acid, boil the solution until its bulk is small and the gas is expelled, add some nitric acid and boil again; and then if the solution is clear, precipitate the iron as peroxide by addition of an excess of aqueous ammonia. Wash, dry, ignite and weigh the precipitate, each 1 part of which equals ·7 part of iron or 3·3475 parts of crystallised green vitriol.

A crude separation of the copper from the iron, &c., may be effected by adding some nitric acid to a small portion of the depositing solution and boiling the mixture for some time, then when it is quite cold adding the strongest liquid ammonia in large excess with constant stirring until all the copper is precipitated and re-dissolved to a clear blue solution. Then separate the clear liquid by decantation and filtration, and wash the residue with solution of ammonia, weaker and weaker until it is quite colourless. Nearly all the copper will then be in the liquid, and all the iron in the solid portion.

In electrolytic refining works, not only the electrolyte but also the batches of crude and refined copper are occasionally analysed, as circumstances dictate.

Purity of the Deposited Copper.—It is a mistake to suppose that electro-deposited copper must be pure; if it has been deposited rapidly from an impure solution it may be very impure.

The purest electro-deposited copper is apt to contain traces of tellurium, bismuth, and silver. These have been detected by photographing the spectrum of the copper. The presence of the two first of these would be due either to the density of the current at the cathode being too great, the liquid insufficiently

stirred, or the proportion of copper being too small, or to all
these causes combined ; but most likely to a combination of the
first and second.   The presence of silver may be due to absence
of soluble chlorides, or to sediment falling from the anodes.

Of all the various circumstances already described which tend
to produce purity of the copper, the most important are :—1.
Absence *in solution* of all the metals which are the most readily
deposited with it : these are silver, bismuth, antimony, arsenic,
and tin ; 2, using a sufficiently moderate density of current at
the cathode ; 3, keeping the solution perfectly clear and well
circulated ; and 4, not allowing any of the mud from the anode
to touch the cathodes.   All other conditions being alike the
most slowly deposited metal is the most pure.

There is usually not much risk of depositing any other metal
than copper (and silver or bismuth if in solution) under the ordinary conditions of working when the density of current does not
exceed 5 amperes per square foot (*see* pp. 208, 209), or the rate
of deposition is not more than 70·7 grains per square foot per
hour, and the increase of thickness is not greater than 1·07
millimetres, or ·04 of an inch per week of 156 hours.   At
Stolberg, with an impure solution, and a density of current of
1·482 ampere per square foot (*see* pp. 208, 235), the purity of
the deposited copper is stated to be "99·92 and 99·95 per cent.
in two analyses."   One source of impurity is fine powder from
the anodes diffused in the liquid.

The difference of purity of the deposit obtained with different
qualities of anodes depends very little upon the electric conductivity of the metal, because with every quality of copper
the large mass of the anode has vastly less resistance than the
electrolyte (*see* p. 25) ; but it depends very greatly upon the
kind of impurities.   If the impurities are soluble and readily
deposited along with the copper, it is then necessary to reduce
the speed of deposition, but this greatly increases the cost.
According to a German writer (Dingler's *Polytechnisches Journal,*
Vol. CCLV., p. 531), the copper deposited at the Oker refinery
is extremely perfect, whilst that of some English electrolytic
refineries is less so ; probably the latter had been too rapidly
deposited (*see* p. 209).   Some samples of American-deposited
copper have also been found to be impure.   A minute proportion
of antimony or bismuth destroys the good qualities of copper.

Chemical Composition of the Mud.—The substances likely to be present in the mud, either frequently or occasionally, are —sulphur, sulphate of lead, silica, carbon, copper in fragments and powder, suboxide of copper, basic sulphates of tin, bismuth, and antimony; gold, silver, platinum, sulphate of lime, sulphides of copper, silver, and iron, chloride of silver, arsenious acid and arseniates ; antimonic acid and antimoniates, particles of slag, &c., and the relative proportions of these ingredients vary with the quality of the anodes, &c.

Maximilian, Duke of Leuchtenberg, was the first to analyse the black sediment which separates from copper when used as an anode in an ordinary sulphate of copper-depositing solution. In 100 parts of the dried substance he found :—

	Per cent.		Per cent.
Tin	33·50	Selenium	1·27
Oxygen	24·82	Gold	·98
Copper	9·24	Cobalt	·86
Antimony	9·22	Vanadium	·64
Arsenic	7·20	Platinum	·44
Silver	4·45	Iron	·30
Sulphur	2·46	Lead	·15
Nickel	2·26		
Silica	1·90		99·69

(*See* Erdmann's "Journal of Practical Chemistry," Vol. XLV., 1848, pp. 460-468.)

This analysis, made forty years ago, clearly shows that, in the process of electro-depositing copper from a solution of its sulphate, the metal is more or less refined and deprived of a large number of impurities, and the minute proportions of precious metals contained in it are concentrated and easily recoverable. The analysis was made purely for the purpose of obtaining scientific knowledge, irrespective of any commercial application ; nevertheless, it has been of great practical value to commercial refiners of copper, and is another instance of the large pecuniary benefit, irrespective of the greater moral one, which, sooner or later, arises from the pursuit of pure truth for its own sake, even when there is no manifest pecuniary advantage to be derived from it ; and the farther knowledge extends, the more are we able to perceive that the history of science is a mass of such examples.

The following analyses of similar residues have been supplied
to me by a friend :—

No. 1.		No. 2.	
Copper	85·85	Lead	27·70
Water and oxygen	4·95	Water and oxygen	21·05
Arsenic	2·48	Copper	19·40
Silver	1·815	Antimony	7·35
Sulphuric acid	1·15	Sulphur	6·35
Insoluble earthy matter	·95	Silver	5·61
Antimony	·75	Arsenic	5·20
Iron	·75	Earthy matter	4·35
Bismuth	·65	Bismuth	1·25
Alumina	·25	Chlorine	·70
Chlorine	·25	Iron	·60
Gold	·085	Nickel	·20
Lead	·05	Organic matter	·20
Loss	·02	Gold	·01
		Loss	·03
	100·00		100·00

Per ton of 20 cwt.

Per ton of 20 cwt.

	oz.	dwts.	grs.		oz.	dwts.	grs.
Silver	623	2	8	Silver	1,835	0	0
Gold	27	15	8	Gold	3	0	0

No. 3.

Copper	67·90
Sulphur	18·10
Iron	5·55
Insoluble earthy matter	3·40
Organic matter	2·25
Lead	2·05
Silver	·55
Loss	·20
	100·00

Effects of the Mud upon the Solution.—Several of the
constituents of the mud are slowly but constantly being dissolved.   Any metallic copper or suboxide of copper in it
gradually oxidises and combines with the free acid to form
cupric sulphate, thus decreasing the amount of free acid and
increasing that of blue vitriol in solution.   The arseniates and
antimoniates are also gradually decomposed by the sulphuric

acid, their bases forming sulphates and their acids being
liberated. The greater the variety and number of ingredients
in the mud, the greater the risk of their being dissolved and
deposited along with the copper. The sulphur, carbon, silica,
gold, platinum (usually also silver), and the sulphates of lead
and of lime, are but little affected. The mud accumulates
sometimes at the rate of several pounds a week in each vat.
In consequence of the action of the liquid upon it, and the
gradual solution of objectionable substances, more particularly
bismuth and antimony, which are rather easily deposited along
with the copper, the mud should not be allowed to remain too
long before it is removed—say three or six months; but this
varies greatly according to the degree of impurity of the crude
copper, and the rate of working. Any dirt upon the anodes is
brushed off in a tank of water and carefully preserved.

Treatment of the Mud.—The baths are emptied of the
clear liquid by means of syphon tubes, the mud taken out, and
washed by mixture with water, subsidence, and decantation of
the clear liquid, which is used in the vats again. The sediment
is then dried. The dry residue is sifted to remove fragments
of copper, and is then either assayed and sold to a metal refiner
or it is melted with litharge or a reducing flux, and the crude
mixture of metals thus obtained is cupelled with some argentiferous lead. Large quantities of silver and gold have been
obtained in this manner. It is stated that as much as 123oz. of
silver and $5\frac{1}{2}$oz. of gold have been thus extracted from $6\frac{3}{4}$ tons
of crude copper; and one writer states that "the quantity of fine
gold gathered at Hamburg in 1880 reached as much as 1,200
kilogrammes " ($=1$ ton $3\frac{1}{2}$ cwt.) The value of such an amount
would be more than £150,000. (*See* " Electrolysis," by Fontaine and Berly, 1885, p. 198.) This is quite possible if the
metal treated was highly auriferous. It was probably derived
from bullion (*see* p. 242).

Cost of Electrolytic Refining.—As the subject of cost belongs
more to the commercial than to the scientific or manufacturing
relations of the process, and therefore does not come within the
intended scope of this book, only a few remarks respecting it
are made.

The cost of refining copper by electrolysis varies with the magnitude of the installation, and depends essentially upon the expense of motive-power, and of interest upon invested capital, also largely upon the degree of "commercial efficiency" of the dynamo, and upon the kind and amount of impurities in the crude metal, which are continually varying, because the composition of each batch of unrefined copper differs. The larger the proportion of foreign metals which are easily deposited with the copper, and the purer the refined copper required, the greater the cost of the process, because it must be conducted more slowly. The cost when anodes composed of the unreduced sulphides of copper and iron are employed instead of metallic copper, as in "Marchese's process" (see p. 229), is considerably different.

Where motive-power or fuel is cheap, advantage is taken of the circumstance to economise the expenditure upon vats and copper, by using it more freely. At Casarza, where water-power is cheap, the proportion of copper deposited is 1·4lbs. per horse-power per hour; at Pembrey, where coal is cheap, it is 2·6lbs.; but at Hamburg, where motive-power and fuel are dear, it is 6·87lbs. The amount of electric energy and of horse-power consumed in depositing the same weight of copper differs in nearly every electrolytic refinery (see pp. 199, 200). Where coal is cheap, and the process is conducted on a large scale, with a good steam-engine, at least 110 indicated horse-power is obtained during one week by the consumption of 50 tons of coal, and with good dynamos will deposit 20 tons of copper per week, at a cost of £20. The interest upon capital invested in plant, &c., during the same period will amount to about a similar sum. Or, if we reckon the cost of one mechanical horse-power at one penny per hour, the cost of such power for electrolytic refining of copper is about one farthing per pound of that metal refined. Difference of potential required to overcome resistance and polarization is also an element in calculating the cost of depositing a given amount of copper, because double the electromotive force expended costs quite double the money.

The chief elements of cost are :—1. Interest upon capital expended upon copper, electrolyte, vats, steam-engine, boiler, dynamo, and other plant. 2. Fuel, water, and oil for the engine.

3. Rent and taxes.   4. Labour.   5. Depreciation of steam-engine, boiler, dynamos, vats, &c.   6. Incidental expenses.

The total stock of copper required in order to refine a given amount of that metal per day or week varies in different works. In each refinery, besides the amount of copper in the vats, there is at least an equal quantity in the form of main conductors, raw copper for making anodes, new anodes ready for immersion, and residues of old anodes, also the stock of refined copper. The value of the stock of copper alone in one of the installations at Hamburg has been estimated at "£8,000." Practically, to refine 30 tons a week of ordinary "Chili bars," with a current density of 8 or 10 amperes per square foot, requires a total stock of about 400 tons of copper, which, at £50 a ton, costs £20,000, to which we must add about £10,000 for plant and premises.

A number of workmen are necessary, probably about twenty, working day and night, when depositing thirty tons of copper per week, to manipulate the electrodes, examine the vats, and attend to the steam-engine and dynamo, melt and cast anodes, wash dirty ones, &c. A chemical analyst is also employed to analyse the copper and the electrolyte. As stoppages in the process do not often occur, they are not a large element of expense; there is, however, the cost of occasional emptying the vats, evaporating and purifying the solution, collecting and treating the mud, &c.

The cost of one of the latest forms of dynamo for electrolytic refining of copper by Messrs. Siemens and Halske, viz., their "H C^{11}" type, constructed to give 1,000 amperes at a difference of potential of 60 volts, was "£420," and that of a 50-unit one, by Messrs. Chamberlain and Hookham, giving 500 amperes at 100 volts, about £400. With large vats, and large currents of low electromotive force, the cost for the same rate of deposition is less, but that of the dynamo and conductors is greater than with small vats and small currents of higher electromotive force. A vat containing one cubic metre costs "about 130 francs," and one of three cubic metres "double that sum." The vats at Biache cost "350 francs each" (p. 181). One cubic metre of the ordinary copper solution costs "about 70 francs."

The total cost of M. Hilarion Roux's installation at Marseilles, including the baths, dynamo, and steam-engine to refine "250 kilogrammes per day" ($=89\frac{1}{2}$ tons yearly), is stated to have been "£1,000," and the amount of invested capital "£5,000 to £6,000." The "Società Anonima Italiana di Miniere di Rame e di Elettro-Metallurgia," in Genoa, whose works are at Casarza, was established with a capital of 4·8 million marks (= £240,000) (Dingler's *Polytechnisches Journal,* 1885, Vol. 255, p. 201).

It might be supposed that the silver and gold recovered constitutes a considerable profit, but all this has to be paid for extra in purchasing the crude metal.

In consequence of the numerous circumstances which affect the cost, an approximately satisfactory estimate of the total amount of capital required to be invested in establishing an electrolytic refinery to deposit a given amount of copper daily, can only be arrived at by means of a knowledge of all the essential particulars in the given case. For further remarks on cost, &c., *see* p. 237.

Recapitulation. — In establishing an electrolytic copper refinery, the chief points usually require to be settled in the following order:—1st. The amount of copper to be deposited per week. 2nd. Amount of mechanical power required and the kind of motor. 3rd. Degree of purity of the solution and of the crude metal. 4th. Rate of deposition per square foot of cathode surface per week. 5th. Total amount of cathode surface necessary. 6th. Total number of vats in series. 7th. Magnitude of each vat. 8th. Electromotive force and strength of current necessary. 9th. Kind of dynamo. 10th. Magnitude of the main conductors. There are, of course, other important points to be first considered, such as the most suitable locality for cheap motive-power, cheap conveyance, &c., but those do not come within the intended scope of this book (*see* also p. 149).

SECTION H.

OTHER APPLICATIONS OF ELECTROLYSIS IN THE SEPARATION AND REFINING OF METALS.

M. Andre's Process for Separating Metals from Coins.— The coins, contained in a basket, form the anode, and a sheet of copper the cathode, in a vat filled with dilute sulphuric acid. Between the electrodes is placed a vertical thin wooden framework, covered on each side with canvas, filled with granules of copper, and dividing the vessel and solution into two portions. On passing the current from a dynamo, the copper and any silver in the coins dissolve, and the liquid diffusing through the diaphragm is deprived of its silver by the copper by the simple contact process (*see* p. 74), and the copper is deposited upon the cathode in the usual manner by the dynamo current. The cathode is made of a conical shape, and, in order to diminish resistance and polarisation, is kept revolving. This process has been worked at Frankfort-on-the-Maine, and has been tried for separating nickel and copper from mattes and speisses, using for that purpose a solution of sulphate of ammonium; and for separating tin from scrap tinned iron by the aid of a solution of common salt.

Keith's Process.—From residuary impure liquors used in making blue vitriol, and containing $4\frac{1}{2}$ per cent. of dissolved copper, with various proportions of silver, nickel, tin, zinc, antimony, and iron, the author recovered "fine merchantable copper," by the ordinary "single cell process" (p. 76). He placed a large porous cell, 32in. deep and 12in. in diameter, in a large oil barrel; the cell was nearly filled with scrap iron and a solution of sulphate of iron free from copper, and the barrel to an equal depth with the cupreous liquid, containing a large cylindrical sheet of copper. The iron and copper were connected together by a wire, and in the course of 36 hours the current generated by the corrosion of the iron deposited the

Q

whole of the copper on the sheet cathode, "as a beautiful velvet-like coat, pure and coherent." For each 56 parts of iron dissolved, an equivalent or 63·5 parts of copper were deposited. By employing 18 of these vessels, the cost was "1 cent per lb. (= £4 per ton of 2,000lbs.) where scrap iron is worth 20 dollars per ton." The cost with a steam-engine and dynamo would have been "14 cents per lb., coal being 8 dollars per ton" (*Proc.* Inst. Civil Eng., 1878, Vol. LII., p. 404).

EXTRACTION OF METALS FROM MINERALS.

The Cementation Process.—The earliest process of separating copper on a large scale from minerals by electro-chemical action is undoubtedly the one in which the drainage water from mines, containing copper in solution derived from the oxidation of mineral sulphides in the earth, is brought into contact with scrap-iron, and its copper deposited by simple immersion. Under such circumstances, the copper separates in little loose crystals, termed "cementation copper," which contains nearly all the impurities of the iron used to precipitate it, and requires to be purified.

For a number of years past has been imported into this country, nearly wholly from Spain and Portugal, by the Rio Tinto, Tharsis, and other companies, about half a million tons annually, of iron pyrites containing several per cent. of copper, the whole of which is extracted by this process. The pyrites, broken into small pieces, is first roasted in a current of air until nearly all the sulphur is burned out for the purpose of making sulphuric acid, and the sulphides converted into oxides. The resulting oxide of iron, containing the copper as oxide and sulphate, is mixed with common salt and again roasted. During this process the copper is rendered more perfectly soluble, and hydrochloric acid vapour mixed with chlorine gas is evolved and condensed. The roasted ore is washed several times with water, and then with the dilute hydrochloric acid, until all the copper salt is extracted. The residuary oxide of iron, termed "purple ore," is used for the manufacture of iron and steel, whilst the copper solution is run into vats containing scrap iron, and all its copper precipitated. The silver is separately thrown down by means of iodide of zinc.

Becquerel's Experiments.—Many years ago, M. Becquerel, sen., made an extensive series of experiments on a commercial scale, extending over ten years, at his works at Grenelle, near Paris, on the electrolytic recovery of silver, copper, and lead from their ores.   The silver ores were converted into chloride, and the lead ones into sulphate, by chemical methods; the chloride and sulphate were then dissolved in saturated solutions of common salt and electrolysed.   The copper ores were converted into sulphate by roasting, &c., and then dissolved in water and electrolysed.   (See *Eléments d'Electrochimie*, 1864, pp. 528–566.)   As dynamos were not then invented, his processes, although essentially good, did not pay.   Some of his methods of chemically preparing the ores for electrolysis have been adopted by various subsequent inventors.

Dechaud and Gaultier's Process.—One of the earliest improvements subsequent to Becquerel's, upon the old cementation method, was that of these inventors, and was introduced in the year 1846.   They converted the mixed sulphides of iron and copper into oxides by roasting them in a current of air. The sulphurous anhydride produced by the roasting of the ore was utilised.   The roasted ore was then washed and dried, and afterwards mixed with sulphate of iron and roasted again so as to convert the copper present into soluble sulphate.   The sulphate of copper was dissolved out by means of water, and the solution concentrated by evaporation.

To separate the copper from this liquid they put the latter into large shallow tanks of wood lined with lead, suspended horizontal perforated plates or grids of iron in the upper part of the solution, and connected them metallically with horizontal sheets of copper or lead (insulated from the vat) in the lower part, and stretched a sheet of canvas between them to prevent the impurities of the iron falling upon the copper.

Under these conditions the iron and copper constituted a voltaic couple; the iron dissolved and caused an electric current which deposited the copper from the liquid upon the sheets. Fresh solution was continually supplied at the upper part of each vat, and the exhausted portion drawn off, and evaporated to separate the sulphate of iron as crystals.   Unfortunately for this process, the immersed iron deposited copper upon itself by "simple immersion" process, and this portion of the copper was contaminated by the whole of the impurities of the iron.

Deligny's Process.—Various other plans have been tried for extracting metals from their native sulphides by the aid of electrolysis. In the year 1881, M. Deligny, reasoning from the fact that the various forms of cupric sulphide containing iron pyrites, &c., are moderately good conductors of electricity, and are more or less quickly attacked by nascent oxygen in the presence of an acid, concluded that if any one of them was used as an anode, the metal in it would dissolve by the aid of the electrical action. He suspended in a weak solution of sulphuric acid a copper plate as a cathode; and as an anode he used pulverised copper ore, contained in a linen bag or porous cell, having in the midst of the ore a piece of carbon for the purpose of making contact. On passing the current from one or two Bunsen cells, the expected action took place.

In one experiment he operated on iron pyrites containing 4·6 per cent. of copper, in which the yellow copper pyrites formed with the iron pyrites a very homogeneous mass. And in another case the copper amounted to 3·6 per cent., and the copper pyrites formed little agglomerations or granules dispersed throughout the mass. After a few days' action of the current both ores were freely attacked, and the copper was partially deposited on the cathode; but the first kind of mineral had all its elements equally dissolved, so that the residue still contained 4·57 per cent. of copper, whilst that of the second contained only 2·35 per cent. "The combination, therefore, of cupreous pyrites with the iron pyrites had been dissolved in a greater proportion than the martial pyrites."

This process does not appear to have been permanently continued, probably because of too great and rapidly increasing conduction resistance at the anode, and incomplete corrosion and solution of the pieces of mineral; and it may be remarked that the method of trusting to loose contacts of the conductor, with a number of pieces of imperfectly conducting substance as the anode, does not succeed. Perfect contact is necessary, especially with inferior conductors such as minerals, and more so with the anode than with the cathode, because as the pieces corrode they acquire a coating of non-conducting impurities, which sooner or later prevent clean contact and stop the current; even when pieces of metal are used in this way, this result occurs after a much longer period. With the cathode this does not take place, because the surfaces of the pieces are kept clean by the continual deposition of metal upon them. With soluble anodes containing non-conducting impurities, even when perfect contact with the conductor is maintained, this practical difficulty is always present, and is greater the larger the proportion of such impurities, and it necessitates a more or less frequent cleaning of their surfaces.

Blas and Miest's Process.—In the year 1882, M. Blas, Professor at the University of Louvain, and M. Miest, an engineer, published a process for separating the metal and sulphur of cupriferous and other mineral sulphides, based upon the facts that those minerals, even when containing a large proportion of gangue, conduct the current with a certain degree of facility, and that when made the anode in a suitable electrolyte, the metal dissolves and the sulphur is left. As an example, they state that with an anode of galena in a bath of nitrate of lead, the lead is dissolved " without the formation of sulphate," and is deposited upon a suitable metal cathode, whilst the sulphur is separated. The solution remains unchanged, and may be used during an indefinitely long period of time. In their process the ore was crushed into grains, about five millimetres diameter, the substance put into a copper mould and subjected to a pressure of about 100 atmospheres; then covered and heated in a furnace to about 600° C.; and finally pressed again, and rapidly cooled in order to more readily separate the plates. The solid mass was used as an anode. When the mineral consisted of zinc blende, the electrolyte was a solution of sulphate or chloride of zinc. Like Deligny's process, this one also has not been carried out permanently on a commercial scale.

Marchese's Process.—This plan was patented in Germany, May, 1882, No. of patent 22,429. The mineral cupreous sulphide in fragments was suspended as the anode, in contact with the conductor, either in baskets immersed in the solution, or else it was placed upon shelves projecting from a carbon

plate as the anode in a solution of a suitable chloride
yielding nascent chlorine by electrolysis, which attacked and
rendered soluble the metals.   The cathode was a sheet of
copper.

The patentee subsequently diminished the electrical resistance of the anodes by compressing the ore in metallic moulds,
and used the compressed plates as anodes.   But even this plan
was not sufficiently successful, and it was found necessary to
previously smelt the ore, so as to diminish the proportion of
non-conducting substances in them and make them more
metallic, also to cast the melted substance into plates in order
to render it more dense and coherent.

The process has been employed in a works at Ponte San
Martino, Piedmont ; and in its improved form is now in
extensive and successful use at those of the Electro-Metallurgical Society of Genoa, at Casarza, Sestri Ponente, near Genoa,
and at the large works of the "Actien Gesellschaft von Stolberg
und Westphalen" at Stolberg, Prussia.   It is essentially a
method for separating pure metallic copper, &c., from a much
more impure material than that usually employed in the
electrolytic refining of copper.   The following is a description
of it as carried out in the works at Casarza, which are arranged
to yield two tons of refined copper per day ; and as the process
is important, I describe it more fully.

The ore, consisting of calco-pyrites mixed with yellow iron
pyrites, and containing a mean proportion of 15 per cent. of
copper, is first smelted without an addition of flux, the serpen-
tine present in it rendering that unnecessary.   It is reduced in
two blast furnaces, "15 tons being smelted in each during 24
hours," with a consumption of "5 tons of 15 per cent." English
coke, and the product is about 5 tons of matte, which is cast
and moulded into about 50 plates, to be used as anodes.   Each
batch of matte is analysed.   The mean of several analyses gave
copper 34·7, iron 38·6, and sulphur 25·3 per cent., as the composition of the anodes.   The iron contained in the matte is
nearly all in the state of protosulphide, as proved by its yielding sulphuretted hydrogen and ferrous sulphate, on addition of
dilute sulphuric acid.   Another portion of the ore, varying in
quantity according to circumstances, is roasted in an ordinary
reverberatory furnace, so as to obtain oxide and sulphate of

copper for the electrolyte, and the evolved sulphurous anhydride is used to produce dilute sulphuric acid.

The waste, which is very great, from the smelting and casting, is reduced to powder in a crusher, and then mixed with hot acid liquid which has been nearly exhausted of copper in the depositing vats. The sulphuretted hydrogen thus produced precipitates the small residue of copper from the solution; it also reduces the ferric sulphate present to ferrous sulphate, and the solution is then either thrown away, or, after having been concentrated by the heat, crystallises out its green vitriol on cooling; the excess of gas is conveyed away by means of a chimney. The ferrous sulphide being thus dissolved out of the powdered matte, and the copper sulphide left, the solid residue is much richer in copper: an analysis of it gave "copper 52·1, iron 18·4, and sulphur 26·3 per cent."

The residue is now dried and roasted in the reverberatory furnace, in such a manner as to convert it into oxides rather than into sulphates, because the iron oxide then formed is very sparingly soluble in dilute sulphuric acid. "The furnace roasts six tons of ore and residue every 24 hours, and consumes 12 per cent. of coal." The sulphurous anhydride from the roasting is converted into dilute sulphuric acid in ordinary leaden chambers, and is used for making the electrolyte, whilst the excess of chamber acid is concentrated into oil of vitriol.

The roasted powder is digested with the dilute sulphuric acid in wide shallow leaden vessels; the oxide of iron remains largely undissolved, whilst the cupric oxide forms a solution of blue vitriol containing "from 3 to 4 per cent. of copper," which is used to supply the depositing vats. The solution is caused to circulate continually, through those vats, at a rate of "about 800 to 1,000 litres per hour for each half-dozen vats," and then over the powder in the leaden vessels, so as to maintain the proportion of dissolved copper in it constant. After the copper is all extracted from the powder, the solid residue is returned to the smelting furnace. Fig. 78 is a diagram plan of the arrangements.

The energy for electrolysing the solution is generated by dynamos of Siemens and Halske's "C 18" type, arranged in two series of ten each. Each armature revolves at a speed of 950 turns a minute, and, with an external resistance of

·0625 ohm, yields 240 amperes at a difference of potential
of 15 volts, and works twelve vats in single series. Water-
power being abundant, the machines are driven by two turbines of 50 horse-power each, and one of 75 horse-power.
Each dynamo with its 12 vats deposits 2 cwt. of copper
in 24 hours, and the total production of refined copper is
two tons daily. This equals 300 amperes, or 75 per cent.

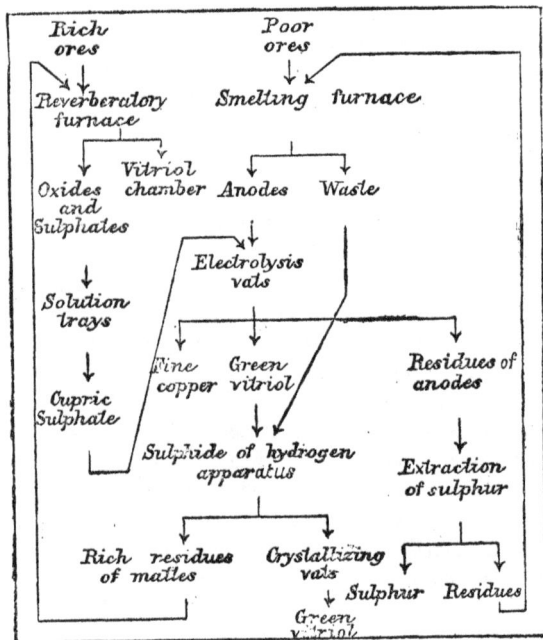

Fig. 78.—Diagram Plan of Marchese's Process.

of effectual current ; some of the excess of current is expended
in reducing persalt of iron.

Each vat contains 15 anodes and 16 cathodes. Each anode
is 32in. long, 32in. wide, $1\frac{1}{4}$in. thick, and weighs 176lbs. The
anodes rest upon two wooden bars fixed upon the bottom of the
vat. Each cathode is 28in. long, 28in. wide, and $\frac{1}{64}$in. thick.
The total amount of active cathode service in each vat is
163·33 square feet, or 19,600 square feet in 12 vats ; and the

density of current is nearly 1·84 ampere per square foot. " For a good circulation the vats must be arranged in the form of a cascade, the fall of each being about ·15 per cent. They are arranged in series of six." Every anode is cast with two strips of sheet copper to serve as connectors, which are inserted in the fused liquid matte at its upper end and become fixed on cooling. Many anodes break during the cooling : the best means of preventing this is by using a matte containing from 20 to 25 per cent. of copper, and cooling the mass very gradually and uniformly, sheltering it from the air.

The anodes contain metallic iron, and consequently acquire a coating of copper when immersed in the electrolyte. The presence of this iron makes them electro-positive to copper, and assists the current, so that only about 1·0 volt per vat has to be provided. In order to protect the suspending strips from corrosion, the upper ends of the anodes are kept about three-quarters of an inch above the liquid. All the sulphur of the anodes is either recovered in the elementary state or is used for making sulphuric acid.

The electrolytic action corrodes the anodes freely, and external layers of impurity (sulphur, oxides, sulphides, &c.), which can be easily separated, gradually accumulate all over their immersed surfaces. Analysis has shown that the outermost layers contain as much as 85 per cent. of sulphur, and are the richest in that ingredient. The separated sulphur, being very porous, does not much increase the resistance. The remains of the anodes are either treated for separation of their sulphur, or they are put into the reverberatory furnace again. The cathodes are changed when the deposit upon them has become " ·5 centimetre " thick : this requires about 12 weeks, equal to a rate of deposition of 1·84 ampere per square foot.

The free sulphuric acid in the electrolyte prevents iron being deposited along with the copper, the proto-salt of iron in solution prevents liberation of oxygen at the anode, and the persalt of iron, of which there is always some present, prevents evolution of hydrogen ; the reduction of this persalt however costs energy. The deposited metal is stated to be " chemically pure."

By washing the roasted matte with dilute sulphuric acid, " a solution was formed containing as much cupric sulphate as was required to render the ferrous sulphide of the anode useful for the electro-deposition of the copper salt." The electroıytꝫ when newly made, contains about 4 to 5 per cent. of copper, and the copper which is continually extracted by electrolysis from it is constantly renewed by the circulation of the liquid over the rich regulus ; but as it also continually takes up iron from that regulus, the sulphate of iron in it gradually accumulates, until, after having been used about two or three months, the quality of the deposited copper begins to deteriorate ; the copper in it is then reduced by deposition to about ·1 per cent., and the liquid removed. The impure liquid contains a large amount of ferrous and ferric sulphate and free sulphuric acid. The residue of copper in it is precipitated by sulphuretted hydrogen, as already described. The mud which settles to the bottom of the electrolysing vats contains sulphur, oxides and sulphides of iron, lead, and copper ; also the silver. If it contains much copper it is roasted over again.

According to Badia, " If the sulphides in the anode contain as much copper as iron, half the current will pass through the copper, and half through the iron (see pp. 92-94). The half that passes through the copper sulphide will operate without any loss, and deposit its equivalent of copper; but the other half, which traverses the iron sulphide, operates at first only upon the copper salt, and finally in reducing the iron persalt : so that, with the same number of coulombs, no more than half the amount of copper is deposited. Consequently, with anodes in which the copper and iron are of equal weight, we may calculate upon a rendering of 75 per cent. of the electric energy expended. If the copper increases, the rendering will be greater, and, on the contrary, less if it diminishes. It cannot, however, get below the 50 per cent. that is obtained when there is no longer any copper in the anodes."

" By employing anodes composed of iron, copper, and sulphur, such as result from a first ordinary fusion of the ore, we may always extract the copper with an electric rendering, which is comprised between 50 per cent., in cases where there is no longer any copper in the anodes, and 100 per cent. when, on the contrary, there is no more iron." " If the baths are properly arranged, and the electrolyte kept at proper strength, a maximum yield of 44 pounds of copper per horse-power may be obtained daily " (see p. 199).

It was found necessary to continually ascertain what variations of potential and strength of current were occurring, in
order to know the conditions of the baths, resistances, &c.
To determine these a Siemens' torsion galvanometer was
employed.

The dimensions and arrangement of the depositing vats,
and the mode of circulating the electrolyte in them, have
been already described (pp. 184, 185, 191, and 195).   The process is best adapted for places where water-power is abundant
and fuel is dear.   For further particulars see a Paper by
G. Badia in *La Lumière Electrique*, 1884, Nos. 40, 41, 42;
also *Scientific American Supplement*, 1885, Vol. XIX., Nos.
478, 479.

Since the establishment of these works the system has been
adopted at the large establishment of the Stolberg and West-
phalian Company at Stolberg, Prussia.   At this place a large
quantity of regulus, containing 7 to 8 per cent. of copper, is
produced in the lead furnaces.   This is calcined and re-smelted
in order to recover as much lead and silver as possible, the
calcination yielding a regulus containing from 15 to 25 per
cent. of copper.

In some experiments made at Stolberg (with the experimental
apparatus used at the International Electrical Exhibition at
Turin), for the purpose of testing the process, the material used
was this regulus : it contained copper 15 to 16, lead 14, iron
41 to 44, and sulphur 25 per cent. ; also 16 ounces 7 penny-
weights of silver per ton.   The electrolyte was the acidulated
wash-water of a richer regulus containing copper 53, lead 14·4,
iron 7·65, and sulphur 15·8 per cent. ; also 18 ounces 6 penny-
weights of silver per ton.   The quantity of copper in the electrolyte varied from 3 to 4 per cent.

The dynamo employed was a Siemens and Halske's " C^6,"
and, running at an average speed of 1,118 revolutions a minute,
gave an average current of 89 amperes at a difference of potential of 5·65 volts, the total external resistance being about
·033 ohm, and the counter-electromotive force in each vat, due
to chemical reaction, being computed to be from ·4 to ·45 volt.
The main conductors were of copper 1·0 centimetre diameter
= ·882 square millimetre per ampere, and there were six vats
in single series.

The vats were of wood, lined with lead, each of the same dimensions, viz.:—

	Length.	Breadth.	Height.
Inside	1,000mm.	680mm.	800mm.
Outside ...	1,160 ,, ...	920 ,, ...	920 ,,

Each contained 7 anodes and 8 cathodes. The anodes were 620 to 640 millimetres long, 600 millimetres broad, and 55 millimetres (usually 30 millimetres) thick; and the cathodes were 600 millimetres long, 600 millimetres wide, and ·5 millimetre thick.

The vats were all on the same level, and were filled to within 10 millimetres of the top of the cathodes, at which level the sulphate solution ran off from each vat into a common gutter to a storage tank, and from that to a clarifying tank, from which a pump lifted it back to a pipe placed opposite to the overflow and supplying the vats. The rate of flow through the vats was 800 to 1,000 litres per hour.

The total amount of copper deposited each 24 hours was 13·322 kilogrammes, or 2·261 kilogrammes per vat = 450 grammes per square metre in 24 hours, or 1·482 ampere per square foot. The theoretical deposit by 89 amperes was 2·5 kilogrammes per vat in 24 hours; the amount actually obtained was "90 to 95 per cent." of this. The deficiency represents energy used in reducing ferric to ferrous sulphate. "A similar percentage of duty has been realised in the works at Casarza, when the solution contained not less than 1 per cent. of copper." The deposited copper was perfectly free from lead and silver, and gave 99·92 and 99·95 per cent. of copper in two analyses (*Proc.* Institution Civil Engineers, 1885, Vol. LXXXII., p. 446).

The plant was worked continuously during two months, and "it is certain that less than 1 horse-power was consumed in the experiments." The dissolved salt of iron continually increased, but did not become so large in amount as to require removal during the two months of working.

"The works (at Stolberg) constructed upon the above data are arranged to produce 10 to 12 cwts. of refined copper a day, corresponding to a consumption of 12 tons of 1st (*i.e.*, 7 to 8 per cent.), or 3 tons of 2nd (15 to 20 per cent.) regulus, and

cover an area of 324 square metres. The number of vats is
58, arranged as series of six in terraces. Each vat contains
about 25 square metres of anodes, and the same surface of
cathodes. The corresponding weights of regulus No. 2 are $2\frac{1}{2}$
tons for each vat, or a total of 145 tons for the 58, representing a value of 14,500 francs (£580) at the works. Of this,
one-half may be considered as being locked up."

"The corresponding lock-up of copper in the cathodes, which
require about three months to acquire a saleable thickness,
may also be taken at one-half, or 45 days at 580 kilogrammes
= 26,100 kilogrammes, worth at present price about 32,000
francs (£1,280). The total value of stock rendered idle will,
therefore, be about £1,600 for a production of 210 tons of
refined copper per annum." "The loss of interest will
(according to M. Marchese) be more than compensated by the
extra price obtained for the product, which is worth from
£5 to £5. 12s. more per ton than the best selected copper"
(*ibid.*, p. 444; Dingler's *Polytechnisches Journal*, 1885, Vol.
CCLV., pp. 199-532; also "Traitement Électrolytique des
Mattes Cuivreuses au Stolberg," par E. Marchese, p. 64:
Genoa, 1885).

Siemens and Halske's Process.—According to this more
recent (1887) method, instead of using soluble anodes of sulphide, the casting of which is attended by great waste of material and labour, and which crumble in the electrolyte and
interfere greatly with the process, insoluble ones of carbon are
employed, with cathodes of sheet copper, and porous diaphragms
between them. In this arrangement the solution of the
sulphates of iron and copper, containing free sulphuric acid,
is caused to flow slowly upward against the cathode, by which
means it is partly deprived of copper, and the persalt of iron in
it partly reduced to ferrous salt ; and then to flow downwards
against the anode, during which time it absorbs the electrolytic
oxygen, and is partly converted into ferric sulphate. The
solution of ferric sulphate thus produced has the property of
converting cupreous sulphide, cupric sulphide, and cupric oxide
into sulphate of copper and rendering them soluble.

In order to again impart sufficient sulphate of copper to
the weakened electrolyte, powdered cupreous pyrites is roasted

at a low temperature, so as to chiefly convert it into cupreous
sulphide and cupric sulphate, and the iron almost entirely into
oxide.   It is then washed in a series of troughs, with the
acidulated and peroxidised liquid from the depositing vats, the
liquid being caused to pass first through the trough containing nearly exhausted powder, and finally through the one last
charged with fresh ore.   During this flow through the troughs
the acid ferric sulphate energetically attacks and dissolves the

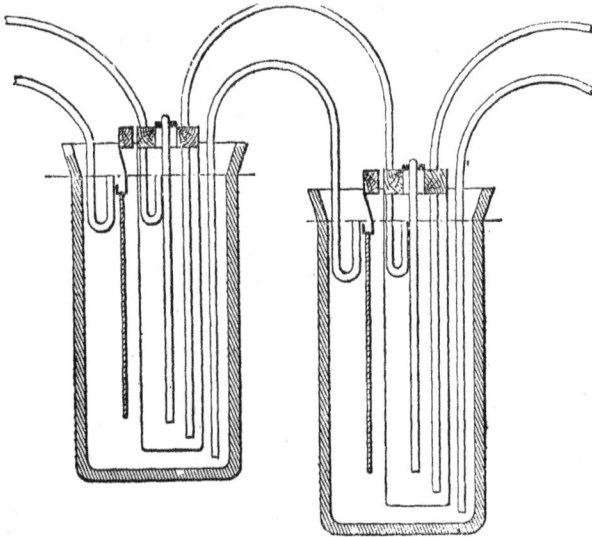

FIG. 79.—Siemens and Halske's Process.

cupreous sulphide, converting it into sulphate, and is itself
reduced to ferrous sulphate.

If the copper in the ore is wholly in the form of cupreous
sulphide, the renovated liquid will contain " exactly the same
amount of sulphate of copper, sulphate of iron, and free
sulphuric acid as before the electrolysis ; but if it exists partly
as oxides, the liquid will be richer in copper, but poorer in iron
and free sulphuric acid."

The processes are based upon the fact that the protosalt of
iron in contact with the anode prevents polarisation at that

electrode by absorbing the electrolytic oxygen and becoming ferric salt. And as the renovated solution contains no ferric salt, none of the electric energy is expended in reducing it to the ferrous state at the cathode.

"As it is generally more convenient to use electric currents of high tension, which require a considerable number of decomposing vats in series, it is necessary to arrange so that the renovated liquid flows consecutively through all the cathode

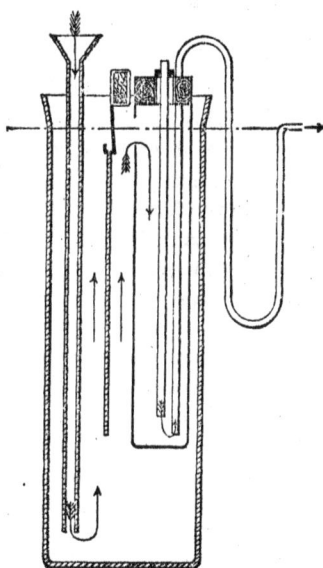

FIG. 80.—Siemens and Halske's Process.

cells, then through all the anode ones, and lastly through the troughs containing the roasted ore." "The same portion of solution is used repeatedly until it becomes too impure for the process."

The vats are arranged in single series and cascade fashion, so that the liquid flows from one vat to another throughout by the action of gravity, as shown in Figs. 79 and 80. The anodes consist of rows of bars of carbon.

ELECTROLYTIC REFINING OF SILVER.

Moebius's Process.—This consists essentially of a mechanical arrangement of brushes for continually keeping the cathodes free from loose crystals of electro-deposited silver, and of muslin bags enclosing the anodes to collect the separated insoluble substances. Beneath the cathodes are trays for catching the silver, and these are lifted out occasionally and the silver removed. The process is specially suitable for copper bullion containing large proportions of silver and gold, with small quantities of lead, platinum, and other metals.

The vats are made of wood, coated inside with graphite paint, and may be arranged either in series or multiple arc as may be desired. The electrolyte consists of dilute nitric acid, containing not more than one per cent. of the acid, and is continually stirred by means of blades hung upon the anode conductors. The bullion anodes are in the form of plates half an inch thick and 14in. square, and the cathodes are made of sheet silver slightly oiled, to prevent adhesion of the deposited metal. By passage of the current, the copper and silver dissolve and form a solution of nitrate of copper, nitrate of silver, and free dilute nitric acid ; the silver alone is electro-deposited as powder and as crystals, leaving the copper in solution; the nitrate of copper is necessary in order to ensure that all the lead is converted into insoluble peroxide at the anode. The peroxide of lead, the gold, platinum, antimony, and some peroxide of silver, separate at the anodes and fall into the bags. The bags are saturated with coal oil, linseed oil, and paraffin, to protect them from the acid, and are very little affected. No porous cells or partitions are employed.

The current must have an electromotive force of one to three volts for each vat. The copper is not deposited provided the liquid is not too poor in silver or too rich in copper ; if a little happens to be deposited, it falls into the trays with the silver, and is then gradually re-dissolved by the liquid. The sediment from the anodes is removed, dried, and melted ; the peroxide of lead then changes to lower oxide, the base metals are oxidised, and the noble ones separate as metal. If platinum or iridium are present, they are subsequently separated by means of bromine.

When too much copper has accumulated in solution, carbon anodes are substituted for the bullion ones, and a feeble current passed until all the silver is deposited. The silver cathodes are then removed, and copper ones substituted. A powerful current is then passed so as to deposit the copper rapidly as a loose powder, which falls into a copper box placed to receive it ; and when so much acid is set free as to corrode the box and its contents, the former is connected as a cathode. The liquid thus regenerated is used again, partly for making new electrolyte, and partly for replacing water evaporated from the baths.

If the base bullion contains as much copper as one-third of the silver in it, it must first be treated as follows :—Plates of the alloy must be used as anodes, and sheets of copper as cathodes, in an acidulated solution of cupric nitrate ; or if the alloy is poor in silver, a solution of cupric sulphate may be employed. By now passing a current of low electromotive force the copper dissolves, and the silver, gold, platinum, &c., remain as a loose coating upon the anodes, the silver being as peroxide, and may be brushed off. This powder must be melted and cast into anodes, and treated in the usual manner.

According to the patentee's statements, " the operation is continuous ; a large quantity of silver can be refined very rapidly, and at a very low cost " (see English patent, Dec. 16, 1884, No. 16,554).

Two dynamos of " C^{18} " type were supplied by Messrs. Siemens and Halske a few years ago for the purpose of working this process, each requiring about 14 horse-power to drive them. The amount of silver refined daily was "about 300 kilogrammes " (Dingler's Polytechnisches Journal, 1885, Vol. CCLV., p. 232).

The efficiency of the process has been proved at Chihuahua, Mexico ; but owing to the large and immediate payments necessary for bullion, the process was stopped. A small plant has been erected at Kansas City, and one by the Pennsylvania Lead Company at Pittsburgh, and 20,000 ounces of bullion refined daily with success, the silver being very pure, " 999 to 999·5 fine," and the cost of the process moderate—about " three-fourths of a cent per ounce." Another has been erected by H. G. Torrey, at No. 153, Cedar-street, New York City, and worked by a Mather-Eddy dynamo, giving a current of 150 amperes with one volt of electromotive force for each vat. The silver obtained was "1,000 fine," and a company is being formed to work the process on a large scale (*American Engineering and Mining Journal*, June 23, 1888, p. 452).

Bullion has also been extensively refined by electrolysis at the works of the North Dutch Refining Company at Hamburg. At these works the separation and purification of gold and silver is successfully practised; "bullion as well as copper is refined. The bullion is cast into plates, and electrolysed by means of a secret solution and method of working." Gold is refined as an anode, "doubtless in a cyanide solution, as in electroplating" (?). The refining of bullion is also carried on at other places. (United States Geological Survey, "Mineral Resources of the United States," 2 vols., by A. Williams, 1883, 1884, pp. 644-649, published at the Government Printing Office, Washington.) The "nitrate" solution said to be employed at Hamburg (*see* p. 146) is probably used in the refining of the bullion somewhat as in Moebius's process just described. At these works a variety of auriferous mattes are refined.

H. R. Cassel's Process.—A method, said to have been brought from Germany (probably Hamburg), where it had been used successfully, of refining gold and silver bullion, not including that containing much lead, was tried in New York during the year 1882 by Messrs. Mathey and Riotte, metallurgists, but did not prove successful, and was soon abandoned. Immediately upon this Mr. Henry R. Cassel, of New York, took out two patents—September 26, 1882—for improvements in the process; a company, termed the United States Bullion Refining Company, was formed to carry out his method, and a plant was erected at 69, Cortlandt-street, New York, to refine 25,000 ounces of bullion per day.

The plant consisted essentially of a 50 horse-power steam engine; a Hochhausen dynamo of extremely low resistance, yielding a potential of 50 volts; a Siemens dynamo of low resistance, yielding 3 volts; twenty wooden tubs, each of 30 gallons capacity, lined with lead, and containing an acidulated solution of cupric sulphate for electrolysing the bullion by means of the Hochhausen dynamo; a series of nine precipitating vats arranged "in cascade" for throwing down the silver by means of immersed copper plates by the "simple immersion process;" and a third series of six vats, containing anodes and cathodes of sheet-lead, for depositing copper out of the solution by means of the Siemens dynamo.

The twenty bullion vats were connected in single series. Six plates of bullion, each 15in. long, 7in. wide, 1in. thick at the top, and ¾in. at the bottom, each weighing about 20lbs., were suspended as anodes in each vat. Between the anodes were placed rectangular porous cells, made of leather, and termed "dialysers." These vessels were erroneously supposed to allow certain of the dissolved substances to pass through them and prevent others. Each "dialyser" contained two sheets of copper to act as cathodes, and dilute sulphuric acid, of specific gravity of 10° Baumé: it was, however, subsequently found that "a certain quantity of nitric acid had been added to the sulphuric acid solution." A very large quantity of hydrogen was evolved at the cathode during the passage of the current, and caused great polarisation and counter-electromotive force ; the leather vessels also offered considerable resistance. The amount of energy consumed in the process by liberating hydrogen, and by the counter current in the bullion vats, and by the polarisation and resistance at the anodes in the six copper-depositing ones, was very large, and "at least 25 horse-power" was consumed in refining the bullion and depositing the copper.

It was assumed that "under the influence of the current, every metal in the anode, except gold," would be dissolved, and that the leather "allowed only the impurities, such as arsenic, antimony, phosphorus, &c., to pass through, whilst preventing the copper." All the dissolved substances, however, gradually passed through, and copper together with silver were deposited instead of hydrogen. The gold set free by the solution of the other metals of the anodes, settled to the bottom as a black sediment, along with "several metalloids" as insoluble impurities.

The acid of the blue vitriol solution in the bullion vats was not allowed to become saturated with the metals of the anode, but was constantly drawn off into the uppermost one of the series of nine precipitating vats, and allowed to run from one to another in succession, during which process the silver was all precipitated, and an equivalent weight of the copper was dissolved. The separated silver was in minute crystals, very pure, and is said to have assayed " ·999 fine."

If the solution in flowing from the lowermost of these vessels did not contain too much copper, it was returned to the bullion vats ; otherwise it was transferred to the six vessels containing the lead electrodes, its copper deposited by the electric current from the Siemens' dynamo, and the liquid then returned to the bullion vats. The copper thus obtained was pure ; any impurities, including the iron, remained in solution.

The chief cause of failure of the process was the presence in the solution of iron dissolved from the bullion : it caused silver to be precipitated along with the gold in the bullion vats, and this rendered necessary an extra refining of the gold (*ibid.*, pp. 646-649).

ELECTROLYTIC REFINING OF LEAD.

Keith's Process.—Patented October 15, 1878, and May 20, 1879. This method has been worked several years, first at 97, Liberty-street, New York, and then by the " Electro-Metal Refining Company," at Rome, Utica, New York State. The arrangement of the vats and electrodes is substantially the same as that adopted in the refining of copper, viz., cast plates of the impure metal are suspended in the liquid from metallic cross-bars, which are connected with the positive conductor of a dynamo-electric machine, and between the plates are similarly suspended thin sheet cathodes of pure lead, connected with the negative conductor. The anodes are enclosed in muslin bags to collect insoluble matter, and are dissolved until only about 2 or 3 per cent. of them is left.

The liquid consists of lead sulphate dissolved in an aqueous solution of acetate of sodium. It is made by electrolysing with lead anodes a mixture of $1\frac{1}{2}$lbs. of acetate of sodium, $2\frac{3}{4}$ ounces of sulphuric acid, and 1 gallon of water, and is heated by steam to about 38°C. The lead of the sulphate is deposited upon the cathode, the sulphuric acid attacks and the acetate dissolves the lead, zinc, and iron of the anodes. Any antimony, silver, or gold contained in the crude metal accumulates undissolved as a soft blue clayey substance in the muslin bags. The zinc and iron, being electro-positive to lead in the liquid, are less easily reduced to metal, and accumulate in solution; and when dissolved in considerable quantity they are to some extent deposited as oxide upon the cathode, which in the subsequent fusion of the deposited metal readily separates. The sediment is collected in the bags, is mixed with borax and nitrate of soda, dried, and melted. The silver separates as metal at the bottom of the crucible, arsenic and antimony are oxidised to the maximum, and unite with the soda to form a scoria. The scoria is digested in hot water to dissolve the arseniate of soda, which is separately crystallised out ; the antimoniate of soda is reduced to metal by carbon and heat, and any iron in it is not reduced, or, if reduced, is removed by fusion with some teroxide of antimony (*ibid.*, Vol. I., 1883, p. 650).

It is stated that in some experiments with 48 wooden vats, each containing 50 anode plates 1·22 metre long, ·38 metre wide, and ·003 metre thick, of base bullion, weighing 16 kilogrammes each, and containing 180 ounces of silver per ton, and 2¼ per cent. of antimony and arsenic, with an expenditure of 12 horse-power, the rate of deposition of lead was 10 tons per 24 hours. The deposited lead contained 11 grains of silver per ton. For each ton deposited the coal consumed was "67·2 kilogrammes." A Weston's dynamo, yielding 1,000 amperes, was employed.

The composition of the lead before and after refining was as follows :—

	Before.		After.	
Lead	96·36	per cent.	99·9	per cent.
Arsenic	1·22	,,	traces.	
Antimony	1·07	,,	,,	
Silver	·5544	,,	·000068	,,
Copper	·315	,,	0·0	,,
Zinc, Iron, &c.	·4886	,,	0·0	,,

The separated lead is said to be pure when the proper conditions have been fulfilled. "The electrolyte is not altered by use."

According to Prof. Barker, of Philadelphia, the cost of the process would not exceed 10 francs per ton, whilst that of the actual treatment of the base bullion by the **dry** method is about 30 francs a ton. Mr. Keith states in his pamphlet that his cost of reduction will be still further reduced.

According to Dr. Hampe, the lead electro-deposited by this process is not perfectly pure, but contains, in particular, a small proportion of bismuth. He electrolysed six litres of a solution of lead acetate acidulated with 4 per cent. of acetic acid, and containing 77·92 grammes of lead per litre. The amount of surface of the electrodes was 13,000 square millimetres. The following table shows the results he obtained :—

	Crude metal.	Deposited metal.	Argentiferous residue.
Lead (by difference) ...	98·79767	99·99297	23·97
Antimony	·55641	·00099	29·70
Copper	·37108	·00060	14·44
Silver	·25400	None	18·435
Nickel	·00730	,,	·090
Tin	·00575	·00041	Traces
Bismuth	·00376	·00305	11·20
Zinc	·00271	·00198	1·80
Sulphur	·00132	None	None
	100·00000	100·00000	99·635

(Dingler's *Polytechnisches Journal*, Vol. CCXLV., p. 515.)

The process was carried on at Rome, U.S.A., on a larger scale as follows :—Thirty circular vats were employed, each being made of a kind of concrete mixture, 6 feet in diameter and 40 inches deep, with a central core or pillar, 2 feet diameter and 40 inches high. Each vat was provided with 13 concentric hoops of sheet brass, 2 feet high and 2 inches apart, to serve as cathodes, with the plates of bullion as anodes suspended between them from a frame of 12 rings of brass, 2 inches wide and ⅛th inch thick, the rings being two inches apart and provided with projecting metal studs, from which to suspend the anodes by means of eye-holes in the suspension lugs. Each frame carried 270 plates of bullion, weighing altogether 2,160 pounds. Each plate was 6 inches wide, 24 inches deep, ⅛th inch thick, and weighed eight pounds. By means of a series of 12 moulds passing in succession under a stream of melted bullion, a man and a boy could cast 180 anodes per hour. The 30 vats were connected in series.

The dynamo employed was one by Edison, giving "2,000 amperes" of current at a potential of "10 volts," and had an internal resistance of "·005 ohm." It was driven by a 10 horse-power steam engine, the speed being such as to cause each anode to dissolve in about 10 days.

A constant flow of solution from an overhead tank was maintained by means of a system of pipes through each vat, and then overflowed through a gutter into a cistern containing 3,000 gallons, from which it was raised again by means of a centrifugal pump to the upper tank, where it was kept at 100° Fahr. by means of steam pipes, and its temperature automatically regulated. The solution was kept neutral. Any iron or zinc in the bullion dissolved in the liquid.

The vats were charged in turn—three each day—with fresh anodes, so that the whole were charged afresh in 10 days; and three tons of bullion were dissolved, refined, and renewed daily.

The deposited lead, which either fell off the cathodes as a soft crystalline layer or was thrown down by scraping, was extracted by opening a plug-hole in the vat, allowing the liquid to run out, and then shovelling out the deposit. The spongy mass of lead was then washed, dried in a centrifugal machine, melted under oil, and cast into ingots.

The process was ultimately abandoned, several years ago, because it would not pay, and the residuary crude metal was purchased by Messrs. Balbach, who refined it by a cheaper method by fusing it with zinc. I am informed that there is now (1888) no company in America refining lead electrolytically.

SEPARATION OF ANTIMONY.

W. Borchers has made some experiments with a view to separating antimony from its mineral compounds on a commercial scale by means of electrolysis. The application is based upon the analytical process of Classen and Lüdwig, and is suitable to any ore or compound of antimony soluble in a solution of sulphide of sodium : the most suitable one is antimonite. The process is applicable to poor ores.

He takes a compound of sodium and sulphur in which those elements are united in equal parts by weight, makes an aqueous solution of it, and dissolves the compound of antimony in the liquid. If the sodium salt contains too much sodium, it offers extra resistance to the current, and if too much sulphur, some of the latter separates during electrolysis. Sequisulphide of antimony, $Sb^2 S^3$, dissolves readily, even in a dilute solution, but the liquid should not be saturated with it : an excess of the sodium sulphide is desirable. So much of antimony sulphide should be dissolved as will bring the solution to a specific gravity of $12°$ Baumé $= 1.091$, and then about 3 per cent. (on the entire solution) of chloride of sodium should be added and dissolved to diminish conduction resistance.

The decomposing vessels and the electrolysis ones may be made of iron, and used as cathodes. The anodes are composed of plates of lead and are not dissolved or peroxidised. An electromotive force of about 2 to $2\frac{1}{2}$ volts per vat is necessary to separate the metal. The decomposition takes place by the electrolysis of three molecules of water for every two atoms of antimony separated. Very nearly all the metal may be deposited out of the solution.

The metal is deposited as powder and as shining scales, some adhering to the cathode : the latter is removed by means of steel-wire brushes. The product is washed, first with water containing a little sulphide of sodium, caustic soda, or ammonia, next with water alone, then with water containing some hydrochloric acid, and finally with water. It is then dried and melted with some glass of antimony (*Chemiker Zeitung*, Vol. II., p. 1,021 ; *Jour*. Soc. Chem. Industry, Vol. VI., p. 673).

SEPARATION OF TIN.

From Scraps of Tinned Iron.—A number of attempts have been made, both by ordinary chemical and by electro-chemical means, to separate the coating of tin from waste scrap and old articles of tinned iron. The following is a brief account of a process for that purpose, described by J. Smith, *Jour*. Soc. Chem. Industry, 1885, Vol. IV., p. 312.

The proportion of tin upon the scraps varied from 3 to 9 per cent., and averaged about 5 per cent. The iron was converted into green vitriol and "iron mordant," and the tin into stannous chloride and other salts used by dyers. The plant was arranged for six tons of scrap per week.

The dynamo employed was a Siemens and Halske's "C^{18}"; it was equal to about half the work required, and was used to commence with. It was driven by a 8 horse-power steam-engine, and was constructed to yield a current of 240 amperes at 15 volts.

There were eight vats, made of wood 5 cm. thick, lined with indiarubber 3·5 mm. thick. Their inside dimensions were 150 cm. long, 100 cm. deep, and 70 cm. wide; and they were placed upon a platform about one metre above the floor. The dynamo was fixed in the adjoining engine-room, and the main conductors passed through a hole in the wall to the vats. Near the vats were two wooden tanks, one on each side the baths, placed to receive the scraps from the electrolytic treatment. A little farther away, and somewhat above the level of the floor, were the dissolving tanks and the evaporating ones. The crystallising vessels were on a still lower level.

The scrap for each electrolysis vat was contained in a strong wooden open-work frame or basket, the inside dimensions of which were 120 cm. long, 85 cm. deep, and 30 cm. wide. It was packed in the baskets with very great care, not loosely nor too closely. Each frame contained 60 to 70 kilogrammes of the scrap, and the entire eight contained about one-half the total quantity required. A number of long and narrow strips of the scrap were distributed vertically in each mass to act as conductors and prevent much heating, their upper ends being all soldered together for convenient attachment to the main conductor. The baskets were emptied of scrap and recharged twice a day.

The cathodes were flat sheets of copper 120 cm. long, 95 cm. wide, and 1·5 mm. thick. There were 16 in all, two in each vat, one being on each side of the basket. Each had a thick rim of copper, square in section to keep it stiff. They were coated with tin to protect them from corrosion. Each cathode was 10 cm. from the side of the basket, and rested in vertical grooves of wood fixed upon the sides of the tank. Each cathode and basket was provided with indiarubber rollers to enable them to be raised and lowered easily without injuring the lining of the vats. There was a travelling pulley above the baths, with which to raise and lower the baskets and plates; also an arrangement of levers and eccentrics fixed to the baths and driven by the steam-engine, to keep the baskets in continual gentle motion in the baths, lifting them 5 cm. about twice a minute.

The electrolyte was composed of a mixture of nine volumes of water and one volume of commercial sulphuric acid of 60° Baume.

The main conductors were cables formed of twisted thick copper wire, enclosed in indiarubber tubing.

The strength of current was measured by means of a Crompton's indicator, which acted up to 250 amperes. With the armature of the dynamo revolving 900 times a minute, the current obtained was close upon 240 amperes. The total resistance of the baths was $= \frac{1}{10}$ ohm.

The tin deposited at first was spongy, but as soon as the acid was partly neutralised, it was thrown down in the form of very minute crystals which fell to the bottom of the bath. It was not quite pure; but when thoroughly washed it was perfectly free from iron.

" The electro-chemical equivalent of tin as a dyad, compared with that of silver, is $\dfrac{117 \cdot 8 \div 2}{107 \cdot 66} = \cdot 546$, and this is equivalent to the deposition of $67 \cdot 65 \times \cdot 546 = 36 \cdot 94$ milligrammes of tin per ampere per minute. For 240 amperes, acting through eight baths in series, we obtain a total deposition of $\dfrac{36 \cdot 94 \times 240 \times 8 \times 60}{1{,}000{,}000}$ $= 4 \cdot 25$ kilogrammes per hour." But little more than half this quantity was, however, obtained, because of " part of the current being absorbed in dissolving the iron as well as the tin, as soon as the former began to get bare "—far more likely, however, because a portion of the current deposited hydrogen in place of tin, the hydrogen being absorbed in reducing some of the oxidised iron salt to the ferrous state. Sulphate of iron formed rapidly, and in about seven weeks the acid was saturated and was run into the sulphate of iron tanks.

During the working of the process, first one and then another of the baths contained the most dissolved iron; but the proportion of dissolved tin remained very constant in each bath and in the total, and averaged 1·5 grammes per litre. Pure tin was deposited until the acid became saturated, and then hydrate of iron began to form.

It was found best, after passing the current during five or six hours, to remove the scrap from the electrolysing to the sulphate of iron tanks, where the iron dissolved with the greatest ease and left the tin, which was recovered and utilised with the remainder.

"One stoker and two or three labourers could by this process work three tons of scrap per week, yielding 3 cwt. of tin. Paris alone could supply 3,000 tons of the raw material yearly, and London a much greater quantity."

It is well known to chemists that a hot aqueous solution of caustic potash or soda, especially with the aid of an electric current, rapidly dissolves tin, but will not dissolve iron.

SEPARATION OF ALUMINIUM.

Kleiner's Process.—A process of separating this metal from cryolite has been invented and patented by Dr. Kleiner, of Zurich. Cryolite is a white mineral, having much the appearance of alabaster or alum. It is a double fluoride of aluminium and sodium, and its composition is represented by the formulæ $3NaF, Al^2F^3$. It contains 54·3 per cent. of fluorine, 32·85 of sodium, and 12·85 of aluminium. It is found in the form of a bed, 80ft. thick and 300ft. long, and of a considerable degree of purity, at Evigtok, in the Arksut Fjord, in West Greenland; also at Miask, in the Ural. It melts at a red heat, and is very slightly soluble in water. It has hitherto been used for making caustic soda, aluminate of soda, to render glass opalescent, and for other purposes.

The process of separating the metal consists essentially in reducing the purest quality of the mineral to powder, immersing in it two rods of carbon in mutual contact at their extremities, passing a copious electric current of high electromotive force, and then slowly separating the ends of the rods until a sufficiency of the powder between them has become fused, and continuing the action for a sufficient time to separate the metal.

The arrangement actually used is as follows :—An iron box is lined with powdered bauxite, rendered compact by being forced in under hydraulic pressure. Through a hole in the bottom of the vessel is introduced a vertical rod of carbon, to act as the cathode. Dry cryolite powder is then placed in the cavity until the cathode is covered. A vertical rod of carbon, supported by a bracket above, and used as a temporary anode, is now lowered into momentary contact with the cathode, so as to produce an electric arc, the electromotive force of the current being about "80 to 100 volts, and the strength 60 to 80 amperes." The anode is then slowly separated, the arc ceases, and the melted substance transmits the current. As the powder melts more is added, and the anode farther separated.

FIG. 81.—Kleiner's Apparatus.

Hollow horizontal cylinders of pure carbon, attached to projecting ears through the sides of the vessel, and on a level with the top of the cathode, constitute the permanent anodes; they are covered with alumina in powder, to prevent their combustion, and are gradually submerged by the melted cryolite. After about ten minutes' action, when sufficient liquid is formed to cover these cylinders, the latter are used as the anode, the temporary one is withdrawn, and the electromotive force is reduced to "50 volts." Gentle fusion is maintained by the current for about three to six hours, and then the process is stopped, and the current switched on to another electrolysing vessel. No fluorine is set free, and no hydrofluoric acid is evolved unless the cryolite powder is damp. Fig. 82 shows the vessel, and Fig. 83 the annular anodes.

By the electrolytic action due to passage of the current, the metal separates at the negative electrode in the form of minute globules, which gradually increase in size to several centimetres diameter, and then fall to the bottom. After sufficient of the metal has been separated, the fused residue is cooled, broken up

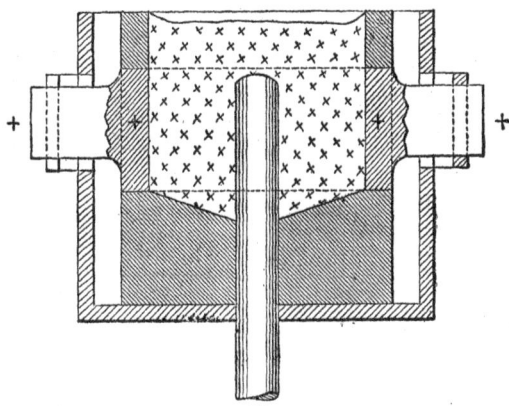

FIG. 82.—Kleiner's Improved Crucible.

and washed, and the metal removed. The soluble residue is preserved for conversion into caustic soda, whilst the insoluble unreduced portion is dried and replaced in the bath.

There are several chief features in this process:—1st. The salt itself is employed alone, and not an aqueous or other solution of it. 2nd. It is reduced to a liquid state by fusion. 3rd. The fusion is effected, not by an external fire or furnace, but by the current itself, i.e., by the heat of conduction resistance. The current therefore performs two functions—it not only electrolyses the saline substance, but also melts it. As the heat is applied only to the interior of the mass and not from the outside, and as the powdered mineral is an inferior conductor of heat, the process may be conducted in a wooden box or other combustible vessel. No more of the substance need be melted than is absolutely required for the electrolysis.

The electrical energy required in the process is considerable, and for the following reasons:—1st. The current has not only to melt the salt as well as electrolyse it, but also to continually make good the loss of heat by radiation and conduction. 2nd. Owing to the powerful chemical affinity of aluminium and fluorine, the electromotive force required to separate the two is

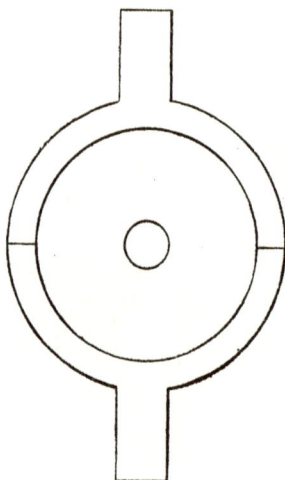

FIG. 83.—Carbon Electrodes.

large. 3rd. As the anode is insoluble the loss of energy at the cathode is not compensated (as in the case of electrolytic purification of copper) by a corresponding gain at the anode; and 4th. The electro-chemical equivalent of aluminium being less than one-third that of copper, the same strength of current separates less than one-third the weight of the former than of the latter metal (see p. 126).

In accordance with this, in an experiment made by Dr. J. Hopkinson, with a mean current of 100·2 amperes, mean potential = 57·43 volts, or a mean energy of 5604·2 watts, during

10380 seconds (= 173 minutes, or 2 hours 58 minutes) = 21·6 horse-hours, the amount of aluminium separated was 60 grammes, the theoretically equivalent amount of the strength of current passed being 93·09 grammes of the metal. From these results " it appears that 3 grammes of metal per horse-power per hour are already attainable when working on a very moderate scale, and a dynamo giving 100 electric horse-power, and working 20 hours a day, would produce 80 pounds of aluminium per week of six days," the theoretical result obtained by an amount of electric energy of one horse-power per hour being 3 grammes, or $\frac{1}{150}$th lb. It has been calculated that the cost of horse-power or mechanical energy required is about " Six shillings and fourpence per pound of aluminium." The cost of purest cryolite is stated to be from £30 to £60 a ton.

Recently (1889) several dynamos, constructed by Messrs. Mather and Platt, Salford, Manchester, and by Messrs. Siemens Brothers, each requiring about 90 horse-power, and yielding 1,000 amperes at 55 volts, and one consuming 60 horse-power, yielding a higher electromotive force, were working at Tyldesley, near Manchester, to test the process on a commercial scale for " The Aluminium Syndicate " of London (see Fig. 61). Patents in connection with the process have been taken out by Major R. Seaver, in England, No. 8,531, June 29, 1886, and by Dr. Kleiner, No. 15,322, November 24, 1886. (See also Jour. Chem. Soc., Vol. VII., pp. 517, 518.)

Heroult's Process.—English patents, No. 7,426, May 21, and 16,853, December 7, 1887. This is a method for producing aluminium and aluminium-bronze, &c., by electrolysis, by pass-ing a continuous and powerful electric current from a dynamo, by means of a vertical carbon bar anode, downwards through alumina melted by the current, either direct into the sides and bottom of a carbon crucible containing the alumina if the metal is required, or into fused metallic copper, and through it into the crucible, when bronze is wanted. The process is com-menced with a small charge of materials.

In this action the alumina is first melted, and then also decomposed by the current, its oxygen appearing at the anode and its metal at the cathode. The oxygen consumes the anode and forms carbonic oxide gas, which escapes, and this action yields an amount of heat which is not necessary to be supplied by the electric current; it also " prevents polarisation, and causes an electric current in the same direction as the principal current, thereby further economising electric energy." " For three equivalents of positive carbon electrode consumed, two equivalents of aluminium are obtained." The anode is lowered in position, and replaced by a new one as quickly as it is consumed, and alumina is supplied as fast as it is reduced.

The carbon crucible is imbedded in a mass of carbon powder within a fire-brick casing, and stands upon a thick plate of carbon, which is connected with the negative pole of the

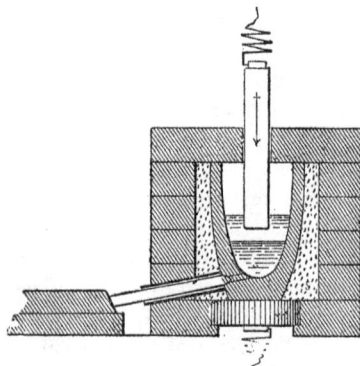

FIG. 84.—Heroult's Process.

dynamo. The production of aluminium is at the rate of " 27 to 29 grammes per horse-power per hour. With a crucible 14 centimetres in diameter a current of 400 amperes is used. The electromotive force of the current is from 20 to 22 volts, which enables more aluminium to be made per horse-power." The melted metal or alloy is run out occasionally through a tapping-hole, provided with a plug in the side and bottom of the crucible. An ampere-meter is included in the circuit to indicate whether the proper distance between the anode and cathode is maintained. Fig. 84 shows the general arrangement.

The process is being carried out on a large scale by the " Schweizerische Metallurgische Gesellschaft " at Neuhausen, by means of Oerlikon dynamos, driven by water-power from the Schaffhausen Falls; and at other places.

The plant in use (June, 1888) consisted of a turbine of 300 horse-power, two dynamos, each giving 12,000 to 13,000

Fig. 85.—Heroult's Apparatus at Neuhausen.

amperes at a potential of 15 to 16 volts, and only one crucible, in which was produced, as required, either aluminium-bronze, silicon-bronze, or ferro-aluminium. In the earlier stage of the

s

process, when aluminium itself was produced, the carbon
crucible was enclosed in a graphite one, surrounded by a fire,

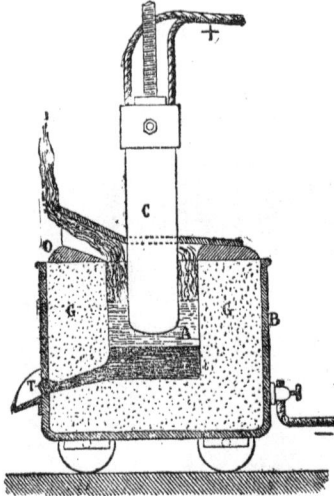

FIG. 86.—Heronlt's Process at Neuhausen.

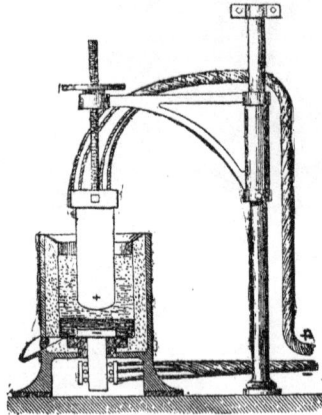

FIG. 87.—Heroult's Process at Neuhausen.

and the alumina was mixed with a large proportion of cryolite
as a flux, to make it more fusible and lighter, and enable the
aluminium to sink to the bottom (see *The Electrician*, 1888,

Vol. XXI., p. 726; *Engineering*, November 30, 1888, p. 540; *Industries*, May 23, 1890, p. 499).

Minet's Process.—English patent No. 10,057, July 18, 1887. In this process the inventor electrolyses a fused mixture of fluoride of aluminium and common salt in a metallic crucible, with carbon electrodes, the cathode resting in a small supplementary crucible placed inside the larger one to receive the separated aluminium. The bath is renewed by occasional additions of alumina (*Engineering*, August 2, 1889, p. 144). A mixture of 40 parts of the double fluoride of aluminium and sodium with 60 parts of chloride of sodium gives the best results (*Nature*, June 19, 1890, p. 192). By this process Minet obtains 30 grammes of aluminium per horse-power per hour (*Chemical News*, July 4, 1890).

Hall's Process.—This consists of electrolysing a fused mixture of the fluoride and oxide of the metal, the latter substance being dissolved in the melted fluoride, and the mixture being contained in a metal crucible having a lining of carbon which forms the cathode. The anode consists of bars of carbon, which are consumed by the oxygen set free by the process "at a rate of nearly a pound of carbon for each pound of aluminium produced." This oxidation of the carbon greatly assists the passage of the current. The mixture preferred is calcium fluoride 234 parts, cryolite 421 parts, aluminium fluoride 845 parts, with small proportions of calcium chloride added occasionally to prevent clogging of the bath. The average E.M.F. of the current is ten volts. The process is employed by "The Pittsburgh Reduction Company," America (*The Electrician* March 28, 1890, p. 515; *Engineering*, March 28, 1890, p. 373; *Scientific American Supplement*, February 1, 1890). The bath conducts about 200 times as well as a sulphate of copper solution. Only the alumina is decomposed by the current. A low red heat is sufficient, and the process is continuous. With an expenditure of 50 mechanical or 45 electrical horse-power, about one pound of aluminium is obtained per hour from each pot. This equals an efficiency of nearly 25 per cent.; about 72 per cent. of the electrical energy is expended in producing the heat. A plant to produce 2,500 pounds of the pure metal is in use at Patricroft, near Manchester (see "Aluminium," by J. W. Richards, 2nd edition, 1890, p. 288).

Diehl's Patent (No. 813, January 16, 1889).—According to this patent, a bath is made of "the fluorides of the alkali-metals, or of the compound fluorides of those metals and anhydrous alum, a sulphate of an alkali and chloride of sodium." This mixture is melted, cooled, ground to powder, and washed thoroughly with water to remove all sulphate. The dried powder is mixed with about as much common salt as is chemically equivalent to the aluminium fluoride present, and melted in a crucible, and the liquid electrolysed with a carbon cathode; or, preferably, with a cathode of iron, copper, &c., when aluminium alloys are wanted (*Jour.* Soc. Chem. Industry, March 31, 1890, p. 397).

According to O. Schmidt, by fusing to a clear liquid, in a well-brasqued iron crucible at a clear red heat, cryolite and common salt, in the proportions indicated by the following equation:—$Al^2F^6 + 6NaF + 6NaCl = Al^2Cl^6 + 12NaF$; and electrolysing the liquid with an anode of gas-carbon and a cathode of copper, the copper did not melt, but acquired a coating of aluminium. The chloride of sodium was not decomposed, nor was metallic sodium separated (*Jour.* Soc. Chem. Industry, Vol. VII., p. 389).

Ludwig Grabau, of Hanover, has patented a process for making fluoride of aluminium, to be used as a substitute for cryolite in the manufacture of aluminium (*Industries*, Oct. 11, 1889, p. 360). Artificial cryolite and fluoride of aluminium are now made by Kempner, in Görlitz, Silesia.

For an account of experiments on the electrolysis of fused cryolite, by W. Hampe, see *Journal* of the Society of Chemical Industry, 1889, Vol. VIII., p. 287. He states that the addition of common salt lowers the fusion point, and enables the aluminium to be separated at a lower temperature (see also *Engineering*, June 21, 1889, p. 693).

At Milwaukee, Wisconsin, U.S. America, "The American Aluminium Company of Milwaukee" has been formed for the purpose of obtaining that metal by the electrolysis of its fused salts. Several other companies have been formed in America for the same object.

P. Marino has patented a process (No. 14,445, October 8, 1888) for electro-depositing aluminium from an aqueous solution (*Jour. Soc. Chem. Industry*, Vol. IX., March 31, 1890, p. 297). L. Brin and A. Brin have also taken out English patents (Nos. 3,547, 3,548, 3,549, and 8,747 of 1888, and 15,508 of October 29, 1889) for the electrolytic separation of aluminium and its alloys (*ibid.*, Vol. IX., January 31, 1890, p. 81). Lossier took out a German patent (No. 31,089) for separating aluminium by the electrolysis of a melted mixture of aluminium fluoride and alkali chloride. Henderson, of Dublin, took out an English patent (No. 7,426) in the year 1887, for using an electrolyte compound of alumina and melted cryolite, with a carbon anode, for obtaining the metal. Messrs. Omholt, Böttiger, and Seidler, of Gossnitz, patented in Germany (No. 34,728) a special form of furnace for electrolysing continuously melted aluminium chloride, and obtaining separately the metal and the chlorine. Feldman's English patent (No. 12,575, September 16, 1887) is for an electrolyte composed of a melted mixture of a double fluoride of aluminium, and an alkali earth metal, with an excess of alkali chloride. L. Grabau patented in Germany (No. 45,012) a melted mixture of cryolite and common salt as the electrolyte. (*See* also A. Winkler's German patent, No. 45,824, May 15, 1888; Menge's ditto, of 1887, No. 40,354; Farmer's English patent of August 6, 1887; and J. W. Richards' book on "Aluminium," 2nd edition, 1890, for modifications and details of the processes for separating the metal from its compounds by means of electrolysis. For Grätzell's apparatus, *see* Fig. 91, p. 268.)

SEPARATION OF ZINC.

Letrange's Process.—This has been tried on a large scale at Saint Denis, and at Romily. According to this method, zinc sulphide ("blende") in small pieces is roasted moderately in a current of air so as to convert as much as possible of it into sulphate, at the same time passing the evolved sulphurous anhydride gas over previously calcined and oxidised ores of zinc, such as calamine or roasted blende, so as to convert them into sulphates, the excess of the sulphurous gas being converted into dilute sulphuric acid in leaden chambers in the usual manner, and the acid subsequently used to dissolve calamine. The zinc sulphate is dissolved in water to form a concentrated solution, from which the metal is deposited by electrolysis, using as anodes either plates of carbon or sheets of lead not attacked by the sulphuric acid, and thin sheet zinc as cathodes.

It is a leading feature of this process that the sulphuric acid used for dissolving the ore is obtained from the ore itself. The preparation of the mineral for electrolysis is said be economical. The roasted ore, or the calamine, which has been converted into sulphate by contact with the sulphurous gas, is quickly converted into sulphate by exposure to the air. A small proportion of blende supplies all the acid required by the lime, magnesia, iron, and other impurities. In order that the

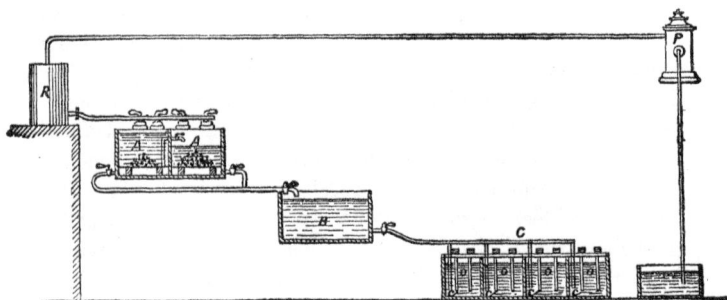

FIG. 88.—Letrange's Process.

roasting may be effectual, the ore is not crushed too finely : after the roasting it is more friable, and is then ground. In order to simplify the process, the sulphite of zinc is dissolved in water, and the solution of that salt is electrolysed, and yields its metal as easily as that of the sulphate.

The roasted and ground ore is placed in large tanks, A (Fig. 88), and a slow current of water allowed to percolate through the mass until all soluble matter is dissolved. The liquor then flows into a reservoir B, and through a series of electrolysis vats C, and a portion of its zinc is deposited as metal by the electric current obtained from a dynamo machine.

The acid set free at the carbon anodes by electrolysis rises and floats in a diluted state upon the top of the zinc solution, and overflows through the openings O into a cistern. It is then raised by means of a pump P into the vessel R, and thence into the washing tanks, containing either calamine or roasted blende, and when saturated with zinc flows back into the electrolysis vats, and thus a constant circulation of the liquid is maintained between the two sets of vessels until the ore is exhausted. As the acid used in forming sulphate of zinc is continually reproduced by electrolysis, all that is lost is that which unites with the lime and other impurities in the ore.

In the electrolysis vats the iron present in the solution is converted into peroxide at the anode, and falls to the bottom as a fine powder. The lead, silver, &c., are insoluble, and also precipitate. The mud is occasionally collected, and its valuable constituents recovered. When the liquid becomes too impure with salts of lime, magnesia, &c., it is removed.

M. Letrange calculated that the cost of installation of a plant for treating one million kilogrammes of zinc by electrolysis would be 500,000fr., or about one-half of that required by the old process; and that by a proper use of the electric current he could attain a daily production of 10 to 12 kilogrammes of zinc per horse-power. In an experiment made in 1882 by M. Cadiat, under the direction of the late M. Alfred Niaudet, with five vats in single series, a current strength of 75 amperes, and an electromotive force of 13·05 volts, continued during four and a-quarter hours, the weight of zinc obtained was 1·475 kilogrammes. A "No. 2" Gramme machine was employed, and had a mechanical efficiency of 75 per cent. The expenditure of work was:

$$\frac{75 \times 13\cdot05}{\cdot75 \times 9\cdot81} = 133 \text{ kilogrammetres.}$$

To obtain 1·475 kilogrammes of zinc, therefore, 133 kilogrammetres had to be consumed during four and a-quarter hours. This equals 565 kilogrammetres per hour, or 5 horse-power for 1 kilogramme of metal per hour (Dingler's *Polytechnisches Journal*, 1882, Vol. CCXLV., p. 455).

"According to M. Kiliani, during the electrolysis of a solution of sulphate of zinc of 1·33 specific gravity, with a zinc plate anode and cathode, the evolution of gas is greatest with a weak current, diminishes as the current gets stronger, and ceases when 3 milligrammes of zinc are deposited per minute upon 1 square centimetre of cathode surface. The precipitate obtained with a strong current was very firm. From a 10 per cent. solution the precipitate was best with a current depositing ·2 to ·4 milligramme of zinc per square centimetre. With very dilute solutions the zinc was always deposited in a spongy state, accompanied by very much hydrogen. With a 1 per cent· solution, and a weak current, oxide of zinc was also precipitated even with an electromotive force of 17 volts, when only ·0755 milligramme of zinc per minute was deposited upon a cathode of 1·0 square centimetre. The size of the electrodes has therefore to be adjusted to the strength of the current and the degree of concentration of the solution" (*Berg und Hüttenmännische Zeitung*, 1883, p. 251; *Jour.* Soc. Chem. Industry, Vol. III., p. 260). Electro-deposited zinc is apt to contain iron.

Lambotte-Doucet's Process.—This has been tried at the Bleyberg mines near Aix-la-Chapelle. The ore is roasted and dissolved in strong hydrochloric acid until the latter is neutralised, and a strong solution of chloride of zinc is obtained. Any dissolved iron is precipitated in the state of ferric oxide by addition of chloride of lime ("bleaching powder") and oxide of zinc. The chloride solution is electrolysed by means of carbon anodes, and thin sheets of zinc as cathodes. As chlorine is liberated at the anode, the solution soon becomes acid ; polarisation occurs at the anode, weakens the current, and zinc ceases to be deposited. The theoretical to the actual loss of energy in electrolysing a solution of chloride of zinc has been calculated to be as "2·7 is to 6·75 horse-power." The yield is " 40 per cent."

Lalande's Process.—This is both a recovery and a refining one. Its essential features are :—1st, the total avoidance of acids, which quickly redissolve the deposited metal; and 2nd, the employment, as a cathode, of mercury, which absorbs the zinc as fast as it is deposited, and thus protects it from contact with the liquid.

The solution consists of zincate of potash and water, and is made by electrolysing caustic potash solution with a zinc anode. It is a very good conductor, suffers very little change of composition by exposure to the air, and may be used a great length of time without requiring to be decarbonated. It will bear quite a dense current at the cathode without evolving much hydrogen, and by using an iron anode, nearly the whole of the zinc may be deposited out of it, and the original alkali regenerated, without much waste of electric energy and without being obliged to employ a very feeble current or occupy much time.

By using a crude zinc anode, the process is suitable for refining impure zinc, the metal deposited being very pure; or, by using an iron one, it may be employed for recovering zinc and regenerating the caustic alkali simultaneously. The deposited zinc is obtained from the amalgam by distilling away the mercury, which may be used over again (see *The Electrician*, August 13, 1886, p. 281). It is in use at Birmingham.

Burghardt's Process.—C. A. Burghardt has patented a process for purifying zinc by electrolysing a solution of the oxide in aqueous caustic soda (*Jour.* Soc. Chem. Industry, July 31, 1880, Vol. VIII., p. 551).

Watt's Process.—This is a patented method (1887, No. 6,294) for separating zinc from its ores, by dissolving the oxide or carbonate in solutions of vegetable acids—preferably acetic acid, containing from 15 to 20 per cent. of real acid; and for depositing the zinc from such solutions by means of electric currents with anodes of carbon, platinum, or crude zinc, the specific gravity of the solution being about 1,150 or 1,160, and the strength of current "three to five amperes."

The patentee states that "when the solutions of zinc are found to contain any considerable proportion of lead, iron, &c., these metals may be precipitated as sulphides by passing a stream of sulphuretted hydrogen through the liquors while in an acid condition." This method is used by chemists to *precipitate* zinc from acetic acid solutions and leave the iron.

Rosing's Process.—Refining of "zinc scum" according to a German patent, No. 33589, October, 1886, granted to the Prussian Government, and used at the lead works at Tarnowitz, in Silesia. The argentiferous zinc lead alloy, containing 45 to 90 per cent. of lead, 8 to 28 of zinc, and 2·5 to 5 of silver, which is obtained when silver is removed by zinc from argentiferous lead by Pattison's process, is finely granulated, then placed in a wooden vat lined at the bottom with lead and filled with a solution of sulphate of zinc. A zinc plate is suspended in the liquid above the alloy, and the lead bottom and sheet of zinc are attached to the two poles of a dynamo, so that the current is upward, the alloy being the anode and the zinc the cathode.

By the electrolytic action, zinc is dissolved from the anode and deposited upon the cathode, but in consequence of the lead being present in large proportion, the action is gradually stopped, much of the zinc still remaining undissolved. The alloy is then removed from the vat, washed, dried, placed in a furnace, and some of the lead removed by liquation. The alloy is then again granulated and returned to the electrolysis vat. The lead which has been removed is sufficiently free from zinc to be cupelled. The processes of electrolysis and liquation are continually alternated, and additional alloy supplied. A French patent has also been taken out for a similar object (Dingler's *Polytechnisches Journal*, Vol. CCLXIII., pp. 87-94 ; *Jour.* Soc. Chem. Industry, 1887, Vol. VI., p. 370).

According to B. Kossman, with one horse-power eight kilogrammes of zinc are deposited by electrolysis in twelve hours. Poor calamines can be treated in this way (*Chemical News,* 1884, Vol. XLIX., p. 69).

For the electrolytic separation of zinc from cadmium, see *Jour.* Soc. Chem. Industry, 1886, Vol. V., p. 41 ; Vol. VIII., p. 639.

SEPARATION OF MAGNESIUM.

Various attempts have been made to separate magnesium by electrolysis on a commercial scale. E. Reichardt, also F. Fischer, employed carnallite and tachyhydrite for this purpose (*see* Dingler's *Polytechnisches Journal*, 1865, Vol. CLXXVI., p. 141 ; 1868, Vol. CLXXXVIII., p. 74; and 1882, Vol. CCXLVI., p. 27). R. Grätzel, of Hemelingen, near Bremen (*ibid.*, 1884, Vol. CCLIII., p. 34), used carnallite and the apparatus shown in Figs. 89, 90 and 91.

In the furnace Q are always, according to the strength of the electric current, two to five melting pots, arranged in a row, each

one in a separate hearth. The crucible-shaped vessel A is composed of cast steel, and forms the negative electrode: it stands

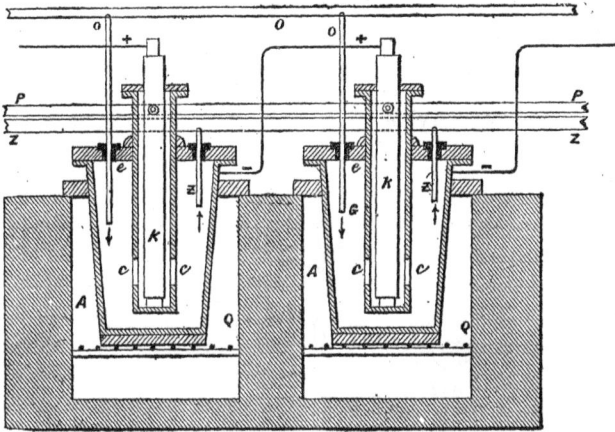

FIG. 89.—Grätzel's Magnesium Process.

upon a refractory fire-brick plate. Each crucible is closed with
a cover, e, of similar material.

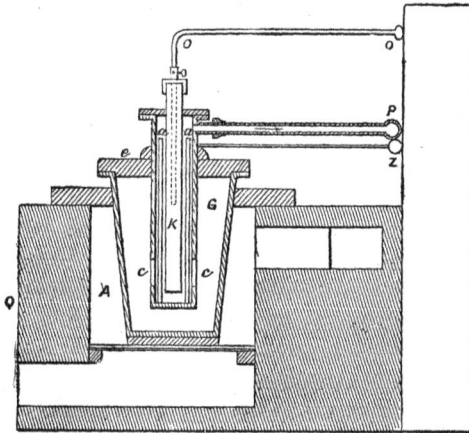

FIG. 90.—Grätzel's Magnesium Process.

The reducing gas arrives through the common supply pipe
O and the tube o into the crucible, and goes back by means of
the tube z in the conduit Z.

The positive carbon electrode k is suspended through an opening in the fire-clay cover of G. The vessel G is cylindrical and formed of fire-clay, which is an electrically insulating substance : it has openings, c, c, on each side for access of the melted substance. Chlorine gas escapes through the common pipe P from each vessel G.

For the recovery of aluminium, the apparatus shown in Fig. 91 is employed : r is connected with the negative electrode.

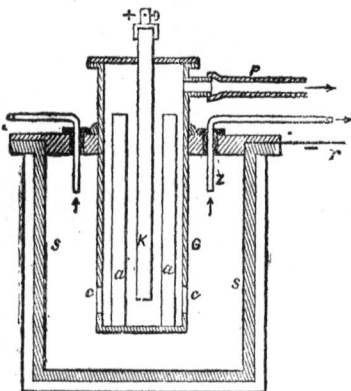

FIG. 91.—Grätzel's Aluminium Process.

For further particulars the reader is referred to the German patent, No. 26962, of October 9, 1883.

Rogers' Process.—A. J. Rogers, of Milwaukee, U.S.A., took out an American patent, No. 296,357, for the separation of magnesium from carnallite. He used sheet-iron crucibles, coated inside with asbestos paste (*see* Fig. 92). *a* and *b* are two cylinders, 13 and 17 centimetres wide respectively, connected below by three strong wires, and supported by three legs. The cover of this furnace is coated on its under side with asbestos paste, and has an opening through which the crucible passes. The bottom of the crucible rests upon thick iron wires. Heat is applied to the crucible by means of three gas flames beneath ; and the products of combustion from them, after heating the crucible, pass down-

wards between the two cylinders in the direction of the arrows
and escape. The material employed for yielding the magnesium was heated to fusion.

When the substance was melted, an annular asbestos plate
was put upon the crucible, and by means of a heavy iron ring, *f*,
fixed by pressure upon the edge of that vessel. The asbestos
plate held a clay tube, *o*, in the side of which small holes were

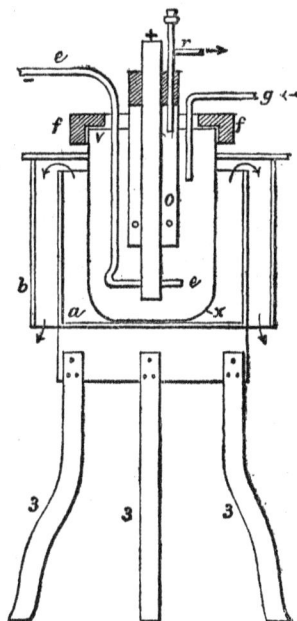

FIG. 92.—Rogers's Process.

bored for the passage of gas. In this tube was supported,
by the help of small asbestos rings, the positive carbon electrode + ; also the joined mouth-pipe *r*, for the escape of
evolved chlorine, having a vertical branch through which the
pipe might be cleared of stoppage. The negative electrode
consisted of an iron wire 5 mm. thick, its lower end being in
the form of a ring surrounding the carbon. Instead of this
ring a carbon plate may be employed.

A current of reducing or indifferent gas, previously dried by means of calcium chloride, was introduced through the tube *g*. This gas passed through the small holes in the clay tube, and together with the chlorine evolved at the anode, escaped by the tube *r*.

A motor of 1 horse-power, and a dynamo giving 50 amperes at 9 to 10 volts, were employed, and 10 grammes of metallic magnesium per hour in a spongy state obtained. Above a red heat balls of magnesium of the size of a hazel-nut were formed, and floated upon the surface a long time.

The patent specification contains also a description of a method for electrolytically separating sodium from its melted chloride, and distilling it out of contact with air (*see* next page).

M. de Monglas has described a process for electro-depositing an alloy of magnesium and zinc from a strong aqueous solution of the mixed chlorides of the two metals, and obtaining the magnesium from the alloy by distilling away the zinc (see *The Electrician*, August 3, 1888, p. 401).

SEPARATION OF POTASSIUM AND SODIUM.

Höpner's Process.—Höpner took out a German patent, No. 30,414, March 21, 1884, for obtaining sodium by electrolysis. Common salt was melted in a crucible containing at the bottom a layer of silver or copper to serve as the anode, connection with which was made by means of a rod of iron or copper. A suitable resisting cathode of metal was suspended in the melted chloride. No precaution appears to have been taken for securing the separated sodium, to prevent its oxidation, or its reunion with the chlorine (Dingler's *Polytechnisches Journal*, 1885, Vol. CCLVI., p. 28).

Jablochoff's Process.—According to this plan the chloride of potassium or of sodium is melted in a covered iron pot, A, over a fire (Fig. 93). Supported by a central hole in the lid is a conical iron funnel, D, with its lower end dipping into the liquid—the funnel being for the purpose of introducing a supply of fragments of the chloride. On opposite sides of the funnel, supported by the lid, are two vertical wide iron pipes, *c* and *c'*, closed at their upper ends, but with their lower ends open and dipping into the liquid. These tubes enclose two vertical electrodes, *a* and *b*, which are supported by the tops of the pipes, and their lower ends dip into the liquid. Exit tubes of iron branch from the upper ends of the pipes, and convey away the evolved chlorine from the positive pole *a*,

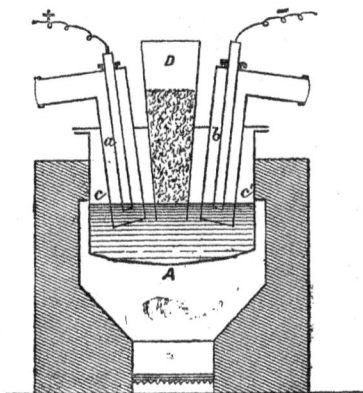

FIG. 93.—Jablochoff's Process.

and vapour of alkali metal from the negative one *b* (Dingler's *Polytechnisches Journal*, Vol. CCLI., p. 422 ; *Jour.* Soc. Chem. Industry, 1884, Vol. III., p. 260).

Rogers's Process.—The apparatus employed consists essentially of three vessels, A, B, and C (Fig. 94). A and B are set in a furnace and heated to redness. A is for melting common salt ; B is for electrolysing the melted substance, and contains two carbon plates as electrodes, separated by a porous partition of unglazed earthenware ; and C is a vessel for condensing the vapour of sodium, which then flows in a liquid state into vessels, D, beneath containing non-volatile oil of petroleum.

A is a cast iron vessel, covered with an air-tight iron lid, which has a funnel for supplying the perfectly dry salt in powder, and is provided with a small valve opening outwards to prevent accidents from explosions. The lower end of the funnel has a hinged valve, which is closed by a floating

metal ball, G, when the liquid rises sufficiently high.   The
melted salt flows through the pipe H and stop-cock I, until
it rises in the vessel B to the same height as in the one A.

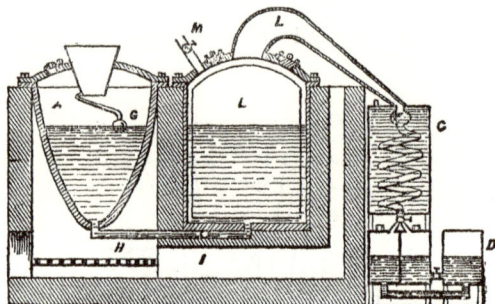

FIG. 94.—Rogers's Process.

B is a rectangular vessel of iron divided vertically into two
equal parts by the porous partition, which extends from the
top or lid nearly to the bottom of the vessel, and is fixed air-
tight at its upper and side edges, so that the vapours rising
from the two electrodes are prevented from mixing.   The
electrodes J, K (Fig. 95), rest upon non-conducting supports of
earthenware, and their conducting wires pass through short

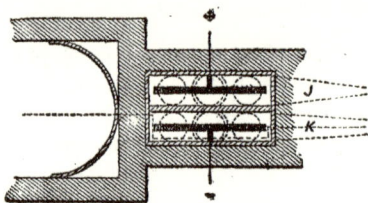

FIG. 95.—Rogers's Process.

fire-clay protecting tubes to the outside of the vessel and
furnace, and on to the dynamo.

Above the two chambers of the vessel, and formed in
one piece with the iron lid, are two bent tubes or retort
necks, L, for conveying away the vapours of chlorine and

sodium respectively. The lid has two manholes, which can be opened when necessary. It has also a pipe and stop-cock M, through which a current of hydrogen can be passed into the chamber containing the cathode until all the air is expelled.

Beketov's Process.—According to the *Moscow Technik*, M. Beketov, of Charkoff, has obtained metallic sodium economically by the electrolysis of common salt. The salt is fused at 500° C., in a horizontal tube of earthenware, and fresh salt is added from time to time through an opening in the middle of the tube. The electrodes are introduced at the opposite ends of the tube, and pass through porcelain tubes which extend beyond the electrodes at their inner ends. The anode is of carbon, and the cathode is a tube of iron through which the liquid sodium continually flows away. "With a current of 16,000 amperes, and an electromotive force of 5 volts (= about 120 horse-power), about 1,800 pounds of salt, yielding 720 pounds of sodium and 1,080 pounds of chlorine, can be decomposed in 24 hours." (*The Electrician*, Vol. XXII., p. 97; *Engineering*, December 14, 1888, p. 588.)

Greenwood's Process.—Mr. James Greenwood, of Queen Victoria-street, London, has recently patented an improved process for the same object, and has used it on a commercial scale for obtaining sodium and chlorine at the works of the "Alliance Aluminium Company," at Wallsend, Newcastle-on-Tyne.

SEPARATION OF GOLD AND SILVER FROM AURIFEROUS EARTH, &c.

Werdermann's Process.—The ores, in a state of fine powder, containing gold, silver, antimony, arsenic, sulphur, &c., are first oxidised by means of ozone, then washed with water, and the silver deposited from the solution by means of electrolysis. The ore is then wetted throughout with a solution of caustic alkali, and the residuary silver, together with the gold, amalgamated with mercury by stirring the mixture with a cathode, the amalgamating vessel being the anode.

T

Barker's Process.—Patented June 28, 1882, No. 3,046, for "abstracting gold and silver from their ores by means of electricity and mercury," especially those containing pyrites, arsenic, black sand, or other heavy substances.

This invention consists essentially of a long and narrow slightly inclined plane of wood, iron, or earthenware, &c., having a series of two or more shallow troughs or riffles formed

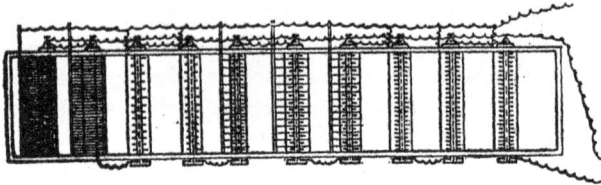

FIG. 96.—Barker's Process.

across its surface at intervals. Each of these troughs contains a layer of mercury, and has a rotating cylinder covered with projecting points, which act as stirrers, above each riffle ; it has also a tap for drawing off the mercury. The speed of rotation of the stirrers is "45 turns a minute." The finely powdered ore is continually supplied at the top of the inclined plane, and constantly washed downwards, by means of water, into and across the mercury in the riffles, and escapes at the bottom.

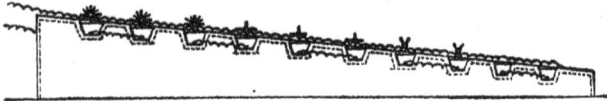

FIG. 97.—Barker's Process.

Each trough is provided with an anode "of brass or any other hard and durable metal," so arranged as to dip into the water, but on no account at any time to touch the mercury which acts as the cathode. The anodes and cathodes are all connected in single series, so that the current from a suitable electric generator passes through all the troughs one after another. Under these circumstances, the mercury is prevented from "sickening," *i.e.*, oxidising, and more readily amalgamates with and absorbs the metals, which are recovered in the usual manner. Figs. 96 and 97 show the general arrangements.

Lambert's Process.—According to this method, into a watertight box, divided by a porous partition, was put a solution of a chloride, and in one division was placed anodes of carbon and the ore, whilst in the other was placed a sheet of copper as the cathode, the mixture of ore and liquid being kept in continual motion by means of a current of water. The gold and silver were deposited upon the copper, and the latter was removed, cleaned, and replaced at intervals.

Molloy's Process.—This inventor has taken out two patents in this subject, No. 143, January 1, 1884, and No. 15,206, November 22, 1886. The latter is an improvement upon the former, and is a mechanical apparatus consisting essentially of a shallow circular tray of iron 36 inches diameter, containing horizontal flat plates of carbon or lead as anodes, immersed in a layer of sand, saturated with dilute sulphuric acid, a solution of caustic alkali or alkaline sulphate. Immediately resting upon this is a circular diaphragm of unglazed earthenware, fitted water-tight at its edges in the vessel. Upon this diaphragm is a layer of mercury about half an inch deep, and floating upon this is a disc of cast iron 35in. wide, with raised edges, capable of being rotated upon the mercury, and having at its centre a hopper-shaped circular opening 9in. wide, through which is constantly supplied a stream of the very finely-powdered ore mixed with water. By revolving the disc about ten or more times a minute, the ore gradually creeps in a spiral direction between the surfaces of the disc and mercury, from the centre to the circumference, in about ten or fifteen seconds, and overflows the edge of the tray into a circular iron gutter. The annexed sketch (Fig. 98) shows the construction of the apparatus.

During the rotation an electric current of about four volts electromotive force (sufficient to overcome the polarisation), from any suitable electric generator, is caused to flow upwards from the anode through the electrolyte, which saturates the sand and the porous diaphragm, into the lower surface of the mercury, causing a deposition of hydrogen and a small amount

of alkali metal into the mercury ; and this hydrogen and alkali
metal reduces the ores of gold and silver to metal, prevents
the "sickening" or oxidising of the mercury, and enables it
to amalgamate with those metals.   The mercury is drawn off
at intervals and distilled away from the gold and silver.
Arrangement is made for the escape of gases evolved at the
anode.   I have not been able to ascertain whether or not the
method is in successful use.

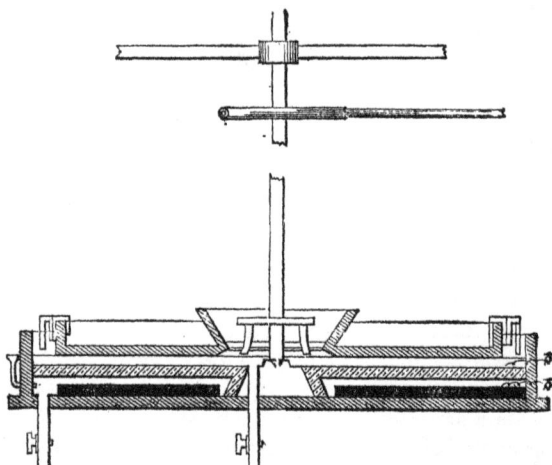

Fig. 98.—Molloy's Process.

H. R. Cassel's Process.— Patented August 9, 1883, No. 3,873 ;
October 13, No. 4,879 ; and July 15, 1885, No. 8,574.   In each
of his patents for separating gold from its ores he generates
chlorine in contact with the finely-powdered ore, by means of
electrolysis of a solution of common salt, at the surfaces of a
number of horizontal carbon rods as anodes (Fig. 99), in a
wooden drum, revolving about eight times a minute upon a
horizontal axis, whilst an electric current is passing through the
mixture.   Under these conditions nascent chlorine and oxygen
are evolved at the carbon surfaces, and come into contact with
every particle of the ore, and rapidly oxidise and dissolve the
gold.   The silver is converted into chloride.   The process is
suitable for "pyrites," ores "containing antimony, arsenic, tellurium, bismuth, &c."

In his second patent, in order to prevent the formation of protochloride of iron, which re-precipitates the gold as metal as fast as it is dissolved, he adds lime or other suitable alkaline

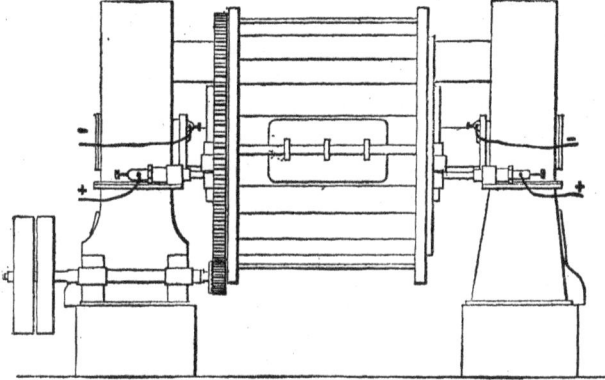

FIG. 99.—Cassel's Process.

earth to the ore, previous to electrolysis ; this also prevents the arsenic and antimony from being dissolved, and obviates the necessity of previously roasting the ore. Chloride of calcium and terchloride of gold are formed at the anode, and caustic soda and metallic gold set free at the cathode.

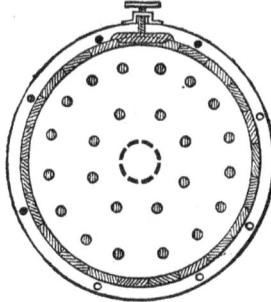

FIG. 100.—Cassel's Process.

In his third patent he electrolytically deposits the gold upon the inner surface of the cylindrical perforated metal axis of the drum, which acts as a cathode, " and subsequently in specially-provided tanks." A mixture of two and a-half tons of the ore with the salt water, is charged into the drum (Fig. 100).

The carbons are all connected together metallically outside the drum, so as to act as a single anode. The current is conveyed to them by means of fixed metallic brushes, pressing against a metal ring, fixed upon the end of the drum, the ring being metallically connected with the carbons.

The axis of the drum is a hollow shaft of iron or copper, through the portion of which inside the drum a number of holes are bored for passage of the liquid and the current, that portion being coated perfectly with rubber varnish to insulate

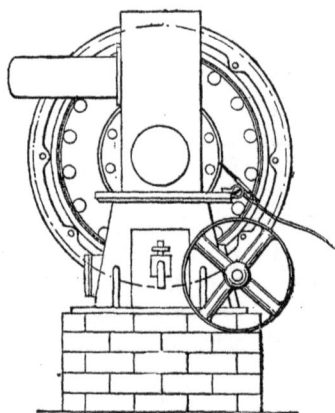

FIG. 101.—Cassel's Process.

it from the liquid, and protect it from corrosion ; it is then covered with a tube of asbestos cloth to act as a filter.

The ends of the shaft pass through stuffing boxes, in metal tanks (Fig. 101), which also serve as supports for the axle. The two tanks are connected by a pipe, to enable the solution to circulate. Provision is made for the escape of gases. The inside surfaces of the shaft and of the tanks act as the cathode. The gold is deposited as a black slime upon the inside of the shaft, and when the drum is revolving, an archimedian screw removes it to the tanks, and causes a circulation of the liquid. The gold is taken from the tanks through movable doors and melted.

When the drum is revolving, the brushes are connected to the positive pole of an electric generator by wire, and the shaft or standards are connected with the negative pole. Several of the apparatuses may be connected together in series in one circuit.

It is stated that "91 per cent. of the gold was extracted from several tons of antimonial concentrates by this process." The process was in use at Glasgow.

Fischer and Weber's Process.—Patented by C. D. Abel, No. 921, January 20, 1887. The ore, ground as finely as possible, is mixed with six or eight times its bulk of water,

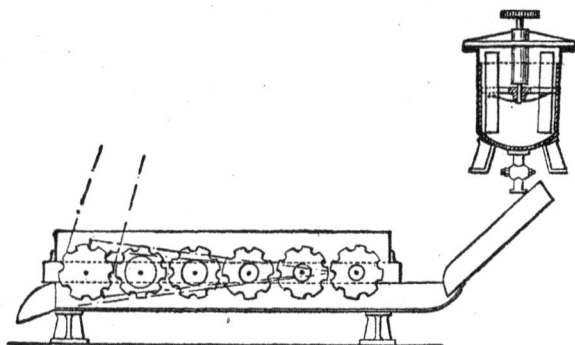

FIG. 102.—Fischer and Weber's Process.

containing, if necessary, dissolved in it, common salt, either alone or in potassium permanganate. The mixture is put into a circular vat with a concave bottom (see Fig. 102); the sides of the vat are partly lined with amalgamated sheet copper, and the concave bottom is covered with mercury touching the copper. It is now agitated during two to two and a-half hours with a rotatory stirring apparatus, having arms, to which are fixed shovels of amalgamated copper connected as anodes with the positive pole of a suitable electric generator, the layer of mercury and the metal lining of the vat being connected as cathodes.

When all the precious metals have been dissolved and deposited, the charge is run off through a large tap in the bottom of the vat into a long horizontal iron trough containing mercury. Above the mercury rotate at different speeds a succession of tranverse wooden rollers, having longitudinal grooves, containing carbon rods as anodes, whilst the mercury is connected as the cathode to an electric generator, and collects any particles of amalgam mixed with the sand. The tailings are then discharged by means of indiarubber brushes or scrapers attached to the last roller (*Jour.* Soc. Chem. Industry, Vol. V., p. 512).

Very similar patents for the same object have also been taken out by A. E. Scott, No. 6,674, June 2, 1885 (*Jour.* Soc. Chem. Industry, Vol. V., p. 431); J. Noad, No. 6,810, May 20, 1886 (*ibid.*, Vol. VI., pp. 516, 517); C. P. Bonnet (American Patent), 1883, No. 298,663; Body (Dingler's *Polytechnisches Journal*, Vol. CCLIV., p. 297); Wiswill (*Engineering and Mining Journal*, 1885, Vol. XXXIX., p. 430), and others.

ELECTROLYTIC REFINING OF NICKEL.

Experiments have been made by Messrs. Siemens and Halske for this purpose; but the electrolytic process required to be repeated, in order to obtain the nickel quite free from iron.

Hermite's Process.—M. Eugene Hermite, of Rouen, has invented a process—said to be practically successful—for extracting nickel from ores, &c. The substance, reduced to fine powder, is digested and agitated in a solution of ammonia, under a pressure of three or four atmospheres, during a quarter to half an hour. The clear solution, after addition of some caustic soda to it, to diminish its conduction resistance, is electrolysed by means of anodes of carbon and cathodes of cast iron in a series of iron vats, provided with sealed or luted hydraulic lids (London *Mining Journal*).

Dynamos have been employed in several nickel works to separate copper from nickel in sulphate solutions. Anodes containing only a small percentage of copper are best suited for the process.

ELECTRIC SMELTING.

Cowles' Process.—This is an invention of Messrs. E. H. and A. H. Cowles, of Cleveland, Ohio, U.S.A., who have taken out

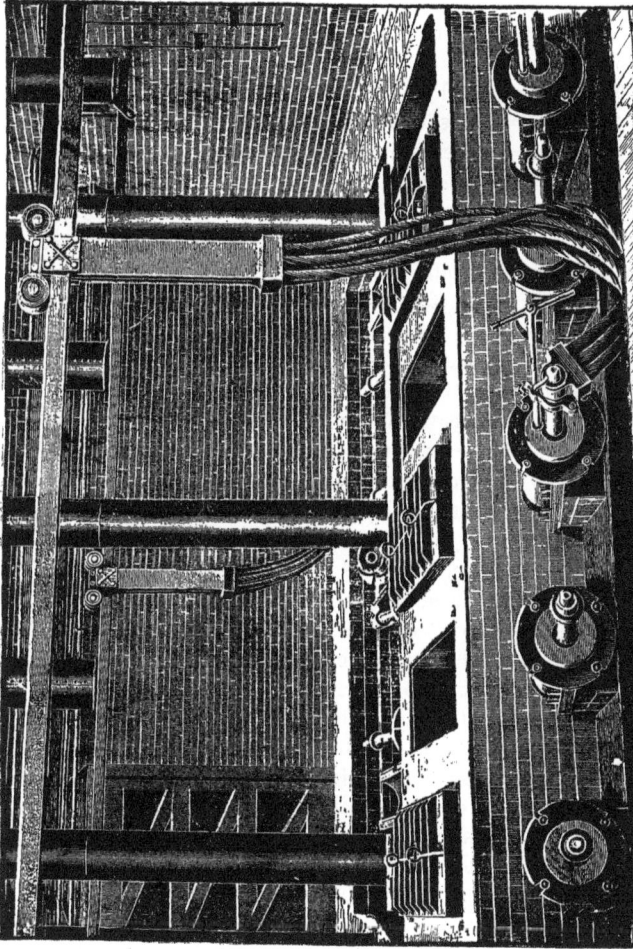

FIG. 103.—General View of the Cowles Furnace.

five patents respecting it in England. It is a method chiefly for producing alloys of aluminium, silicon, and other difficultly-reducible metals, with copper, iron, &c., and less for separating

or refining those elements by means of electrolysis. It is
largely a chemical action, in which carbon, at the intensely
high temperature of the electric arc, instead of that of an
ordinary furnace, and higher even than that of the oxy-
hydrogen blow-pipe, reduces various oxides to metal; but
it appears also to be influenced to some extent by electrolysis.

The dynamo employed for producing the current and heat is
of the Brush kind, shunt-wound (*see* p. 165), yielding a direct
current, and is kept quite away—in a separate apartment—
from the dust and grit of the furnace-room. Its weight is
more than 7,000lb., and at a speed of 907 revolutions a
minute yields a current of 1,575 amperes, at a difference of
potential of 46·7 volts.

The main conductor is composed of 13 copper wires, each
·3in. diameter; it includes in circuit an ammeter, through
which the entire current flows, indicating by its attraction of a
piston armature, attached to a spring balance and an indicator
dial, the total strength of current. The face of the indicator is
in full view of the workers at the furnaces. It will take from
50 to 2,000 amperes, and indicate them on the dial.

In the furnace-room, between the meter and the furnaces,
and in the main circuit, is placed a large resistance coil of
German silver wire, immersed in water, any portion or the whole
of which can be either short-circuited or thrown into the circuit
by means of a heavy copper slide, which may be readily moved
to or fro. It is thrown into circuit whilst the current is transferred from one furnace to another, or when it is necessary to
gradually weaken the current before entirely stopping it. It is also
thrown into action when a short circuit occurs in the furnace
The annexed sketches show the construction of the furnace and
the arrangement of its contents (Figs. 104, 105, 106).

The furnace is simply a rectangular box, its inside dimensions being 5ft. long, 1ft. wide, and 15in. deep; it is built of
fire-brick. It has a movable iron lid, lined with fire-brick,
which entirely screens it; the lid has three holes in it for
the escape of gases. At opposite ends, sliding through copper
boxes filled with copper shot, and supported by porcelain wheels,
are the electrodes, composed of cylindrical bars of prepared
hard carbon, about 3in. diameter and 30in. long : thicker ones

are apt to disintegrate. In larger furnaces as many as five pairs of carbons 4ft. long are used.

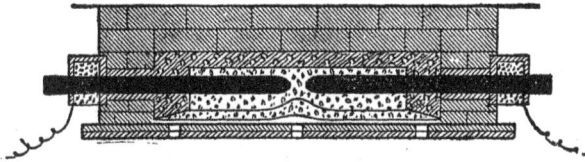

FIG. 104.—Cowles' Process.

The furnace is charged as follows :—The electrodes are inserted with their ends touching each other. Finely-powdered charcoal is wetted with lime-water and dried to coat it with a

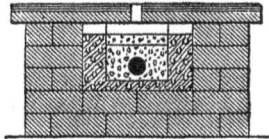

FIG. 105.—Cowles' Process.

film of lime : if unlimed charcoal is used it gradually becomes changed into graphite, and then conducts away the heat. The bottom of the furnace is lined to a depth of two or three inches

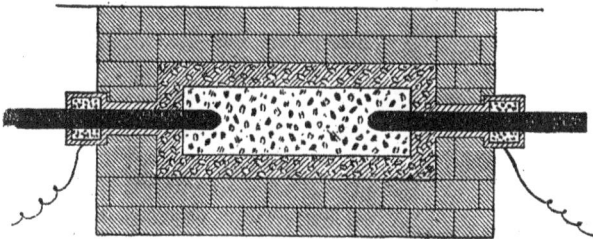

FIG. 106.—Cowles' Process.

with this powder. A sheet-iron rectangular gauge, like the four sides of a box, nearly 12in. deep, is placed upon the powder, leaving about two inches all round between it and the walls of the furnace, and this outer interval on all sides is filled with the limed charcoal. The charge, composed of a mixture of about 25lb. of colourless corundum (crystallised native oxide of aluminium), or bauxite, in bits about a quarter of an inch diameter, 12lb. of charcoal dust and small bits of charcoal, and 50lb. of granulated copper, is placed beneath, around, and above each electrode, and in the entire rectangular space until it is filled to about 9in. deep. The remaining central space is filled with small fragments of charcoal, and the gauge is then taken out. This upper layer of coarse charcoal allows the gases to escape. The cover is now placed upon the furnace, and the crevice luted to exclude air.

To commence the action the connections are made with the copper boxes and dynamo, and sufficient resistance is switched into the circuit to make it safe to start the machine. The latter is started, the carbons slightly separated, and the ammeter continually watched ; the wire resistance is gradually lessened, and at intervals the carbons drawn farther apart, until in about ten minutes the copper between the electrodes has all melted, and the electrodes have been moved so far apart that the current has become steady. The current is now allowed to increase by diminishing the wire resistance until about 1,300 amperes at 50 volts is passing.

The furnace is now fully at work, and carbonic oxide gas is escaping and burning at the holes in the cover. The action is continued about five hours, the internal resistance of the furnace being kept constant by slight movements to and fro of the electrodes. This regulation has been made self-acting, the electrodes being moved by means of a shunt circuit, electro-magnet and vibrating armature. The cooling influence of the copper shot and boxes prevent the red-hot parts of the electrodes being burned by the air. At the end of the reduction the electrodes project considerably out of the furnace ; the current is reduced by inserting the wire resistance, and then switched on to another furnace.

In this furnace, under the influence of the heat of conduction resistance, generated by a current of 65,000 watts of electric energy, or 87 electrical horse-power, corundum is reduced to aluminium by carbon in the absence of copper, and a yellow carbide of aluminium is formed, and amorphous aluminium set free. It is stated that the oxides of boron, silicon, titanium, chromium, magnesium, sodium, and potassium, have all been reduced to metal by carbon in this furnace. Numerous alloys of different metals with these have also been made. It is difficult to work the process without an excess of carbon (*Jour.* Soc. Chem. Industry, Vol. V., p. 331). For more complete information respecting this process, see *The Electrician*, Vol. XXI., p. 589 ; *Jour.* Soc. Chem. Industry, September 30, 1889, pp. 677-684 ; *The Chemical News*, November 1, 1889, p. 211. ; *Industries*, Sept. 7, 1888, p. 237.

According to W. Hampe, the reduction of alumina to metal by means of carbon, even at the temperature of the electric furnace is, on thermo-chemical data, impossible ; and the effect of the electric current in the Cowles process, is in the first place electro-thermic in melting the alumina by the heat of electric conduction-resistance, and then electrolytic in decomposing the fused substance. He electrolysed a melted mixture of cryolite, common salt, and chloride of calcium with a carbon anode and a cathode of fused copper. Chlorine separated at the anode, and sodium at the cathode. The sodium rose to the top of the melted mixture in fused globules and burned, but scarcely a trace of aluminium was separated (*Jour.* Soc. Chem. Industry, Vol. VII., p. 236).

It has been stated that the radiation of heat and light from the electric furnaces at Creusot blinds the workmen, and causes the skin to peel off their necks and faces.

APPENDIX.

Decimal Equivalents of Inches and Feet.

Fractions of an inch.		Decimals of an inch.		Decimals of a foot.	Fractions of an inch.		Decimals of an inch.		Decimals of a foot.
$\frac{1}{16}$	=	·0625	=	·00521	$\frac{9}{16}$	=	·5625	=	·04688
$\frac{1}{8}$	=	·125	=	·01041	$\frac{5}{8}$	=	·625	=	·05208
$\frac{3}{16}$	=	·1875	=	·01562	$\frac{11}{16}$	=	·6875	=	·05729
$\frac{1}{4}$	=	·25	=	·02083	$\frac{3}{4}$	=	·75	=	·06250
$\frac{5}{16}$	=	·3125	=	·02604	$\frac{13}{16}$	=	·8125	=	·06771
$\frac{3}{8}$	=	·375	=	·03125	$\frac{7}{8}$	=	·875	=	·07291
$\frac{7}{16}$	=	·4375	=	·03645	$\frac{15}{16}$	=	·9375	=	·07812
$\frac{1}{2}$	=	·5	=	·04166	1	=1·0		=	·08333

Area of Circles.

Diam.	Area.	Diam.	Area.	Diam.	Area.	Diam.	Area.
·1	·007854	1·625	2·074	4 25	14·186	7·25	41·261
·125	·012246	1·75	2·404	4·5	15·9	7·5	44·156
·25	·049	1·92	2·894	4·75	17·72	7·75	47·147
·375	·11	2·	3·14	5·	19·63	8·	50·266
·5	·196	2·25	3·974	5·25	21·649	8·25	53·456
·625	·306	2·5	4·906	5·5	23·76	8·5	56·745
·75	·4418	2·75	5·939	5·75	25·967	8·75	60·099
·875	·6013	3·	7·668	6·	28·274	9·	63·617
1·	·7854	3·25	8·295	6·25	30·63	9·25	67·2
1·1	·95	3·5	9·62	6·5	33·183	9·5	70·846
1·25	1·227	3·75	11·045	6·75	35·785	9·75	74·622
1·5	1·767	4·	12·566	7·	38·485	10·	78·540

To find the area of a circle, multiply the diameter by itself, and the product by ·7854.

English Measures of Capacity.

—	Cubic Inches.	Pints.	Fluid Ounces.	Fluid Drachms.	Minims.
Imperial Gallon .	277·276	8	160	1,280	76,800
,, Quart .	69·318	2	40	320	19,200
,, Pint ...	34·659	...	20	160	9,600
Fluid Ounce	1·733	...	...	8	480
Fluid Drachm ...	...	...	...	...	60

One cubic foot equals 6·232 gallons or 1,728 cubic inches.

One cubic yard equals 168·264 gallons.

English Weights.

AVOIRDUPOIS.

—	Grains.	Drachms.	Ounces.	Pounds.
Drachm	27·3			
Ounce	437·5	16		
Pound	7,000·	256	16	
Quarter............	196,000·	7,168	448	28
Hundredweight.	784,000·	28,672	1,792	112
Ton	15,680,000·	573,440	35,840	2,240

—	Grains.	Ounces.	Pounds.
Fluid drachm of water	54·7		
,, ounce ,, ...	437·5		
Imperial pint ,, ...	8,750·	20	1·25
,, quart ,, ...	17,500·	40	2·5
,, gallon ,, ...	70,000·	160	10·
One cubic inch ,, ...	252·5		

TROY.

—	Grains.	Pennyweights.	Ounces.
Pennyweight	24		
Ounce	480	20	
Pound	5,760	240	12

French Measures.

Length.			Weight.		
		Millimetres.			Milligrammes.
Millimetre			Milligramme		
Centimetre	equals	10	Centigramme	equals	10
Decimetre	,,	100	Decigramme	,,	100
Metre	,,	1,000	Gramme	,,	1,000
Kilometre	,,	1,000 metres	Decagramme	,,	10,000
			Hectogramme	,,	100,000
			Kilogramme	,,	1,000,000
			(equals 1,000 grammes)		

Capacity.

Millilitre equals 1· Gramme, or 1 cubic centimetre of water at 4°C.
Centilitre ,,   1· Decagramme.
Decilitre ,,   1· Hectogramme.
Litre   ,,   1· Kilogramme = 1 cubic decimetre of water at 4°C.
  ,,   ,, 1000· Grammes.

Conversion of French and English Measures.

Capacity.

Cubic Centimetre............	equals	·061027	cubic inches.	
,, Metre	,,	35·36	,, feet.	
Litre	,,	61·027	,, inches.	
,, ...	,,	35·2155	fluid ounces.	
,, ...	,,	1·76077	pints.	
,, ...	,,	·2201	gallon.	
Cubic Inch	,,	16·386	cubic centimetres	
Fluid Ounce	,,	28·3966	,, ,,	
Pint	,,	·5679	litres.	
Gallon	,,	4·54346	,,	
1 Cubic Metre	,,	220·41	gallons.	

Weight.

Milligramme	equals	·01543	grains.
Gramme	,,	15·432	,,
Kilogramme....................	,,	15,432·	,,
,, ...	,,	35·274	ounces.
,, ...	,,	2·2046	pounds.
Litre..............................	,,	15,432·	grains.
Grain.............................	,,	·0648	grammes.
Ounce	,,	28·3495	,,
Pound	,,	·45359	kilogrammes.
Hundredweight	,,	50·8024	,,

Length.

Millimetre .... equals	·03939in.
Centimetre .... ,,	·3939in.
Metre.................. ,,	3·2809ft. = 3ft. 3in. and ⅓rd.
Kilometre .... ,,	1093·6 yards.
Inch .... ,,	25·3995 millimetres.
Foot .... ,,	·304795 metre.

Surface.

1 square metre = 10·78 square feet = 1,552·36 square inches.
1 ,, foot = ·0929 ,, metre = 92,900 ,, millimetres.
1 ,, inch = 645· ,, millimetres.
1 circle 1·0 inch diameter = 506·7 square millimetres.
1 ,, ·1 ,, ,, = 126·677 ,, ,,
1 ,, 1·0 centimetre = ·12186 square inch.

Specific Gravities of Liquids corresponding to degrees of Baume's Hydrometer. (Poggiale.)

Deg.	Sp. Gr.	Deg.	Sp. Gr.	Deg.	Sp. Gr.	Deg.	Sp. Gr.	Deg.	Sp. Gr.
0	1000	15	1116	30	1264	45	1453	59	1691
1	1007	16	1125	31	1275	46	1468	60	1711
2	1014	17	1134	32	1286	47	1483	61	1732
3	1022	18	1143	33	1297	48	1498	62	1753
4	1029	19	1152	34	1309	49	1514	63	1774
5	1036	20	1161	35	1320	50	1530	64	1796
6	1044	21	1171	36	1332	51	1546	65	1819
7	1052	22	1180	37	1345	52	1563	66	1846
8	1060	23	1190	38	1357	53	1580	67	1872
9	1067	24	1199	39	1370	54	1597	68	1897
10	1075	25	1210	40	1383	55	1615	69	1921
11	1083	26	1221	41	1397	56	1634	70	1946
12	1091	27	1231	42	1410	57	1652	71	1974
13	1100	28	1242	43	1424	58	1671	72	2000
14	1108	29	1253	44	1438				

Twaddell's Hydrometer.—To convert degrees of this hydrometer into specific gravities, multiply them by 5 and add 1,000.

Specific Gravities of Metals.

Metal.	Sp. Gr.	Metal.	Sp. Gr.	Metal.	Sp. Gr.
Platinum ...	21·53	Cobalt.........	8·95	Tellurium ...	6·25
Iridium	21·15	Copper	8·95	Arsenic	5·97
Gold	19·34	Nickel.........	8·82	Aluminium...	2·60
Mercury	13·60	Cadmium ...	8·70	Strontium ...	2·54
Thallium ...	11·90	Manganese...	8·01	Magnesium ..	1·75
Palladium ...	11·80	Iron............	7·84	Calcium	1·58
Lead	11·36	Tin	7·29	Sodium	·97
Silver	10·53	Zinc............	7·15	Potassium ...	·87
Bismuth	9·80	Antimony ...	6·71	Lithium	·59

One cubic inch of Copper weighs 2257·35 grains.

Relations of Thermometric Scales.

9 Fahrenheit degrees = 5 Centigrade degrees = 4 Reaumur degrees.

To convert

Fahrenheit to Centigrade, subtract 32, multiply by 5, and divide by 9.
      „   „   Reaumur,   „   32,   „   „ 4, „   „   „ 9.
Centigrade to Fahrenheit, multiply by 9, divide by 5, and add 32.
      „   „   Reaumur,   „   „ 4,   „   „ 5.
Reaumur to Fahrenheit,   „   „ 9,   „   „ 4, and add 32.
      „   „   Centigrade,   „   „ 5,   „   „ 4.

Example: 212° Fahrenheit to Centigrade, $212 - 32 = 180 \times 5 \div 9 =$
100° Centigrade.

Fusibility of Metals.

Metal.	Fusibility Deg. C.	Metal.	Fusibility Deg. C.
Mercury	− 39·4	Aluminium ......⎱	Above a red
Potassium	+62·5	Calcium............⎰	heat
Sodium	97·6	Silver	1023·
Lithium	180·	Copper	1091·
Tin	228·	Gold................	1102·
Cadmium............	228·	Cast iron............	1503·
Bismuth	264·	Cobalt ....⎫	
Thallium	294·	Nickel	The highest
Lead	325·	Manganese ......⎬	heat
Zinc	412·	Silicon	of a forge.
Antimony about...	620·	Wrought iron ...⎭	
Magnesium a little		Palladium ......⎫	Require the
below	800·	Platinum .... ⎬	oxyhydrogen
		Iridium .... ⎭	flame.

Table of Centigrade and Fahrenheit Degrees.

Deg. C.	Deg. F	Deg. C.	Deg. F.	Deg. C.	Deg. F.	Deg. C.	Deg. F.	Deg. C.	Deg. F.
0	32·	21	69·8	41	105·8	61	141·8	81	177·8
1	33·8	22	71·6	42	107·6	62	143·6	82	179·6
2	35·6	23	73·4	43	109·4	63	145·4	83	181·4
3	37·4	24	75·2	44	111·2	64	147·2	84	183·2
4	39·2	25	77·	45	113·	65	149·	85	185·
5	41·	26	78·8	46	114·8	66	150·8	86	186·8
6	42·8	27	80·6	47	116·6	67	152·6	87	188·6
7	44·6	28	82·4	48	118·4	68	154·4	88	190·4
8	46·4	29	84·2	49	120·2	69	156·2	89	192·2
9	48·2	30	86·	50	122·	70	158·	90	194·
10	50·	31	87·8	51	123·8	71	159·8	91	195·8
11	51·8	32	89·6	52	125·6	72	161·6	92	197·6
12	53·6	33	91·4	53	127·4	73	163·4	93	199·4
13	55·4	34	93·2	54	129·2	74	165·2	94	201·2
14	57·2	35	95·	55	131·	75	167·	95	203·
15	59·	36	96·8	56	132·8	76	168·8	96	204·8
16	60·8	37	98·6	57	134·6	77	170·6	97	206·6
17	62·6	38	100·4	58	136·4	78	172·4	98	208·4
18	64·4	39	102·2	59	138·2	79	174·2	99	210·2
19	66·2	40	104·	60	140.	80	176·	100	212·
20	68·								

USEFUL DATA.

Mechanical Units.

Units
Of time ... ... = 1 minute.
,, space ... ... = 1 foot.
,, force ... ... = The force which will support 1 pound weight.
,, work ... ... = 1 foot-pound, or the quantity of energy which will raise 1 pound through a space of 1 foot.
,, rate of work ... = 1 horse-power, or 550 pounds per second, = 33,000 per minute.
French unit of work ... = 1 kilogrammetre, 1 kilogramme raised through 1 metre, = 7·233 foot-pounds, = 9·807 joules (p. 37).

2

Electrical Units.

Units

Of quantity of electric   = 1 coulomb (p. 12).
    current

  ,, rate of flow of elec-   = 1 ampere, = 1 coulomb per second
    tric current   (p. 18).

  ,, conduction-resistance   = 1 ohm, or that of a column of mer-
                                 cury 1·0 square m.m. section and
                                 1,060 m.m. long at 0° C. (p. 24).

  ,, electromotive force   = 1 volt, or that which will urge 1 cou-
                                 lomb through a resistance of 1
                                 ohm in 1 second, = ·9268 that of
                                 a Daniell's cell, = 22,900 calories
                                 (pp. 14-49).

  ,, electric energy   ...   = 1 watt or volt-ampere, = $\frac{1}{746}$ horse-
                                 power (p. 23), = $\frac{1}{736}$ French horse-
                                 power.

  ,, electric work   ...   = 1 joule or volt-coulomb = ·7375 foot-
                                 pounds (see Notes, p. 37).

  ,, electrolytic work ...   = The quantity of water, viz., ·000093
                                 gramme, decomposed by one cou-
                                 lomb of current (p. 128).

Thermal Units.

Units

Of quantity of heat   ...   = 1 calorie, = 1 gramme of water raised
                                 1° C. = 423·55 grammes raised 1
                                 metre high, = 3·0636 foot-pounds.
                                 English unit, 1 pound of water
                                 raised 1° Fahr., = 772 foot-pounds
                                 "the mechanical equivalent of
                                 heat" (p. 37), = 252·0 C. G.
                                 calories.

  ,, heat of electric con-   = 1 joule, or the heat produced by 1
    duction resistance   coulomb passing through a resist-
                                 ance of 1 ohm, = 1 watt exerted
                                 during 1 second, = ·24 of a centi-
                                 grade-gramme calorie (p. 37).

Chemical Unit.

Unit

Of chemical energy   ...   = The energy exerted in the formation
                                 of ·000093 gramme of water (p. 38).

MISCELLANEOUS DATA.

1 foot-pound = 1·356 Joules.
1 watt = 44·236 foot-pounds per minute, or ·7373 per second.
1 horse-power = 746 watts = 76 kilogrammetres per second.
  unit of electric output = 1,000 watts = 1·34 horse-power.
  horse-power = 746 amperes = 1·93 pounds of copper deposited per
  hour (pp. 19-28).

1· square foot of copper 1·0 mm. thick, weighs 29·25 ounces.
1·   „   „   „   „   ·01 inch „   „   7·43 „
1·   „   „   „   „   ·25 inch „   „   11·6 pounds.
1·   „   „   „   „   weighing 1·0lb. = ·0208 mm., or ·0008
                    inch in thickness.
1· cubic inch „   „   weighs 5·1585 avoirdupois ounces, or
                    ·3224 pound.

1·0 ampere deposits 1·174 gramme or 18·116 grains of copper
                    per hour (pp. 19-128).
1·0   „   „   4·0248 grammes or 62·100 grains of silver
                    per hour (pp. 19-128).
10·0   „   „   9·84 ounces of copper in 24 hours.
112·7   „   „   1·0 pound of silver per hour.
248·5   „   „   1·0 kilogramme of silver per hour.
386·4   „   „   1·0 pound of copper   „   „
746·0   „   „   1·93 „   „   „   „   „
851·8   „   „   1·0 kilogramme of copper „   „

7·87 amperes per square foot deposit ·25 inch thickness of copper
  per four weeks of 156 hours per week.
17·94 amperes per square foot deposit ·001 inch thickness of
  copper per hour.
69·96 amperes per square foot deposit ·10mm. thickness of copper
  per hour, or 15·6mm. per week.

179·4 amperes per square foot deposit ·01mm. thickness of copper per hour.

⅜ths oz. of copper deposited per square foot per hour = $\frac{1}{12}$ inch increase of thickness per hour.

1·0 lb. of copper deposited per square foot per hour = ·547mm. increase of thickness per hour.

The absolute zero of heat and of electric conduction resistance in metals is = – 273° C., or – 459·4 Fahr. (p. 31).

INDEX TO CONTENTS.